Selected Titles in This Series

178 **V. Kreinovich and G. Mints, Editors,** Problems of Reducing the Exhaustive Search

177 **R. L. Dobrushin, R. A. Minlos, M. A. Shubin, and A. M. Vershik, Editors,** Topics in Statistical and Theoretical Physics (F. A. Berezin Memorial Volume)

176 **E. V. Shikin, Editor,** Some Questions of Differential Geometry in the Large

175 **R. L. Dobrushin, R. A. Minlos, M. A. Shubin, and A. M. Vershik, Editors,** Contemporary Mathematical Physics (F. A. Berezin Memorial Volume)

174 **A. A. Bolibruch, A. S. Merkur′ev, and N. Yu. Netsvetaev, Editors,** Mathematics in St. Petersburg

173 **V. Kharlamov, A. Korchagin, G. Polotovskiĭ, and O. Viro, Editors,** Topology of Real Algebraic Varieties and Related Topics

172 **K. Nomizu, Editor,** Selected Papers on Number Theory and Algebraic Geometry

171 **L. A. Bunimovich, B. M. Gurevich, and Ya. B. Pesin, Editors,** Sinai's Moscow Seminar on Dynamical Systems

170 **S. P. Novikov, Editor,** Topics in Topology and Mathematical Physics

169 **S. G. Gindikin and E. B. Vinberg, Editors,** Lie Groups and Lie Algebras: E. B. Dynkin's Seminar

168 **V. V. Kozlov, Editor,** Dynamical Systems in Classical Mechanics

167 **V. V. Lychagin, Editor,** The Interplay between Differential Geometry and Differential Equations

166 **O. A. Ladyzhenskaya, Editor,** Proceedings of the St. Petersburg Mathematical Society, Volume III

165 **Yu. Ilyashenko and S. Yakovenko, Editors,** Concerning the Hilbert 16th Problem

164 **N. N. Uraltseva, Editor,** Nonlinear Evolution Equations

163 **L. A. Bokut′, M. Hazewinkel, and Yu. G. Reshetnyak, Editors,** Third Siberian School "Algebra and Analysis"

162 **S. G. Gindikin, Editor,** Applied Problems of Radon Transform

161 **K. Nomizu, Editor,** Selected Papers on Analysis, Probability, and Statistics

160 **K. Nomizu, Editor,** Selected Papers on Number Theory, Algebraic Geometry, and Differential Geometry

159 **O. A. Ladyzhenskaya, Editor,** Proceedings of the St. Petersburg Mathematical Society, Volume II

158 **A. K. Kelmans, Editor,** Selected Topics in Discrete Mathematics: Proceedings of the Moscow Discrete Mathematics Seminar 1972–1990

157 **M. Sh. Birman, Editor,** Wave Propagation. Scattering Theory

156 **V. N. Gerasimov, N. G. Nesterenko, and A. I. Valitskas,** Three Papers on Algebras and Their Representations

155 **O. A. Ladyzhenskaya and A. M. Vershik, Editors,** Proceedings of the St. Petersburg Mathematical Society, Volume I

154 **V. A. Artamonov et al.,** Selected Papers in K-Theory

153 **S. G. Gindikin, Editor,** Singularity Theory and Some Problems of Functional Analysis

152 **H. Draškovičová et al.,** Ordered Sets and Lattices II

151 **I. A. Aleksandrov, L. A. Bokut′, and Yu. G. Reshetnyak, Editors,** Second Siberian Winter School "Algebra and Analysis"

150 **S. G. Gindikin, Editor,** Spectral Theory of Operators

149 **V. S. Afraĭmovich et al.,** Thirteen Papers in Algebra, Functional Analysis, Topology, and Probability, Translated from the Russian

148 **A. D. Aleksandrov, O. V. Belegradek, L. A. Bokut′, and Yu. L. Ershov, Editors,** First Siberian Winter School "Algebra and Analysis"

(Continued in the back of this publication)

American Mathematical Society

TRANSLATIONS

Series 2 • Volume 178

Problems of Reducing the Exhaustive Search

V. Kreinovich
G. Mints
Editors

American Mathematical Society
Providence, Rhode Island

EDITORIAL COMMITTEE

AMS Subcommittee
Robert D. MacPherson
Grigorii A. Margulis
James D. Stasheff (Chair)
ASL Subcommittee Steffen Lempp (Chair)
IMS Subcommittee Mark I. Freidlin (Chair)

V. Kreinovich and G. Mints (editors)

ПРОБЛЕМЫ СОКРАЩЕНИЯ ПЕРЕБОРА

Научный Совет по Комплексной Проблеме
«Кибернетика»
Академия Наук СССР
Москва, 1987

Translated by V. Minachin

1991 *Mathematics Subject Classification*. Primary 68Q15; Secondary 03D15, 68Q60.

ABSTRACT. This collection contains translations of papers on propositional satisfiability and related logical problems. A significant role in the collection is played by papers related to Maslov's iterative method of search reduction. The book is useful for researchers and graduate students working in mathematical logic, theory of algorithms, and in related areas of computer science.

Library of Congress Cataloging-in-Publication Data

Problemy sokrashcheniia perebora. English.
  Problems of reducing the exhaustive search / V. Kreinovich, G. Mints, editors.
    p. cm. — (American Mathematical Society translations ; ser. 2, v. 178)
  Includes bibliographical references.
  ISBN 0-8218-0386-7 (alk. paper)
  1. Computational complexity. I. Kreinovich, Vladik. II. Mints, G. E. III. Series.
QA3.A572 ser. 2. vol. 178
[QA267.7]
510 s—dc20
[511.3]
                                                                                96-27577
                                                                                  CIP

Copying and reprinting. Material in this book may be reproduced by any means for educational and scientific purposes without fee or permission with the exception of reproduction by services that collect fees for delivery of documents and provided that the customary acknowledgment of the source is given. This consent does not extend to other kinds of copying for general distribution, for advertising or promotional purposes, or for resale. Requests for permission for commercial use of material should be addressed to the Assistant to the Publisher, American Mathematical Society, P. O. Box 6248, Providence, Rhode Island 02940-6248. Requests can also be made by e-mail to reprint-permission@ams.org.
  Excluded from these provisions is material in articles for which the author holds copyright. In such cases, requests for permission to use or reprint should be addressed directly to the author(s). (Copyright ownership is indicated in the notice in the lower right-hand corner of the first page of each article.)

© 1997 by the American Mathematical Society. All rights reserved.
The American Mathematical Society retains all rights
except those granted to the United States Government.
Printed in the United States of America.

∞ The paper used in this book is acid-free and falls within the guidelines
established to ensure permanence and durability.

10 9 8 7 6 5 4 3 2 1   02 01 00 99 98 97

To the Memory of
Sergei Maslov and Nina Maslova

Contents

Foreword to the English Translation	ix
Algorithmics of NP-Hard Problems R. I. Freidzon	1
Algorithmics of Propositional Satisfiability Problems E. Ya. Dantsin	5
Semantics of S. Yu. Maslov's Iterative Method V. Ya. Kreinovich	23
Ergodic Properties of Maslov's Iterative Method M. I. Zakharevich	53
Anomalous Properties of Maslov's Iterative Method I. V. Melichev	65
Possible Nontraditional Methods of Establishing Unsatisfiability of Propositional Formulas Yu. V. Matiyasevich	75
Dual Algorithms in Discrete Optimization G. V. Davydov and I. M. Davydova	79
Models, Methods, and Modes for the Synthesis of Program Schemes A. Ya. Dikovsky	103
Effective Calculi as a Technique for Search Reduction M. I. Kanovich	133
Lower Bounds of Combinatorial Complexity for Exponential Search Reduction N. K. Kossovskiĭ	149
On a Class of Polynomial Systems of Equations Following from the Formula for Total Probability and Possibilities for Eliminating Search in Solving Them An. A. Muchnik	161
S. Maslov's Iterative Method: 15 Years Later (Freedom of Choice, Neural Networks, Numerical Optimization, Uncertainty Reasoning, and Chemical Computing) V. Kreinovich	175

Foreword to the English Translation

This collection contains translations of papers on propositional satisfiability and related problems in logic which appeared in a book "Problemy Sokrashcheniya Perebora," Voprosy Kibernetiki no. 133, published in Russian in 1987 by the Scientific Council "Cybernetics" of the USSR Academy of Sciences. These problems form the nucleus of an intensively developing area of investigation. A general description of this area and of the separate papers is given in a short introductory note by R. Freidzon. Some of the later developments are described in a supplement written by V. Kreinovich for this translation. Related recent investigations involving ideas and results of the present volume concern such diverse topics as biological computing [1], combinatorial optimization [2], decision problems for linear logic [5], improving upper bounds for satisfiability algorithms [6], and applied logic programming [3].

This translation is dedicated to the memory of two remarkable Russian mathematicians, Sergei Maslov [9] and his wife Nina Maslova.

S. Maslov is known as the author of the inverse method in automated deduction [7], which was discovered at the same time as the resolution method of J. A. Robinson and has approximately the same range of applications. In 1981, Maslov proposed an iterative algorithm for propositional satisfiability based on some general search ideas described in detail in his book [8], published posthumously. When in 1982 Maslov was killed at the age of 43 in a car accident, a seminar was started to follow through on his ideas. The result of this collective research forms the bulk of the present book.

Nina Maslova was a mathematician in her own right. She was awarded the National Prize in Mathematics, one of the highest distinctions for mathematics in the former Soviet Union. When she died in 1993 at the age of 54, the St. Petersburg Mathematical Society convened a special meeting in her honor.

The intellectual influence of the Maslov family was not restricted however to their scientific achievements. Their home in Leningrad (now St. Petersburg) was a meeting place of a seminar where talks on social and scientific problems were presented. One has to feel the gravity of the ideological pressure of a totalitarian state to appreciate the importance of such a free forum. The emergence of such seminars seems to be characteristic for intellectual life under oppressive regimes: recall Zilsel's seminar in Vienna where Gödel presented in January 1938 his overview [4] of possibilities for continuing Hilbert's program. Another forum for dissident thought in the USSR was provided by a Samizdat (unofficially published) journal

"Summa" edited by S. Maslov [10], which was designed as a review journal for Samizdat publications.

One may hope that this book will contribute to further progress of knowledge and understanding, which was the goal of Nina Maslova and Sergei Maslov.

Grigori Mints   Stanford, June 1996

References

1. L. Adleman, *Molecular computation of solutions to combinatorial problems*, Science **266** (1994), 1021–1024.
2. G. Davydov and I. Davydova, *Tautologies and positive solvability of linear homogeneous systems*, Ann. Pure Appl. Logic **57** (1992), 27–43.
3. A. Dikovsky, *On computational complexity of Prolog programs*, Theoret. Comput. Sci. **119** (1993), 63–102.
4. K. Gödel, *Lecture at Zilsel's*, Collected Works of Kurt Gödel (S. Feferman et al., eds.), Vol. III, Oxford University Press, 1995, pp. 87–113.
5. M. Kanovich, *The complexity of neutrals in linear logic*, Proc. 10-th Annual IEEE Symposium on Logic in Computer Science (San Diego, California, 1995), pp. 486-495.
6. O. Kullmann and H. Luckhardt, *Various complexity upper bounds for decisions on propositional tautology*, Inform. and Comput. (1996) (to appear).
7. S. Yu. Maslov, *An inverse method of establishing deducibility in the classical predicate calculus*, Dokl. Akad. Nauk SSSR **159** (1964), 17–20; English transl., Soviet Math. Dokl. **5** (1964), 1420–1424.
8. _____, *Theory of Deductive Sytems and Its Applications*, "Radio i Svyaz", Moscow, 1986; English transl., MIT Press, Cambridge, MA, 1987.
9. G. Davydov, Yu. Matiyasevich, G. Mints, V. Orevkov, A. Slisenko, and N. Shanin, *Sergei Yuryevich Maslov. Obituary*, Uspekhi Mat. Nauk **39** (1984), no. 2 (236), 129–130; English transl., Russian Math. Surveys **39** (1984), no. 2, 133–135.
10. A. Vershik, *Concealed digest of stagnation times*, Zvezda **1991**, no. 11. (Russian)

Algorithmics of NP-Hard Problems

R. I. Freidzon

ABSTRACT. The article surveys main directions of research in the area of constructing efficient methods for solving NP-hard problems. Recent results on Maslov's iterative method of search reduction are presented.
   A brief review of the subsequent articles in the book is given.

1. The intensive research done in recent decades on the algorithmics of NP-hard problems was inspired not only by the deep interest of research mathematicians in solving the fundamental theoretical problem $P = NP$ but also by acute practical needs. The notion of a universal search problem (or NP-complete problem) has come to symbolize the difficulties encountered by algorithm designers in a wide range of important applied problems. Problems of this kind often arise in computer aided design, planning and production management systems, distributed computations, information and expert systems, and many other fields. This list is rapidly growing while the problem of eliminating exponential-time search remains open. As is often the case in the relationship between theoretical and applied research, an elegant mathematical theory of NP-complete problems constructed in the early 1970s serves only to produce a systematic record of success and failure in our efforts to solve search problems. An extensive literature is devoted to the algorithmics of NP-hard problems.

A brilliant discussion of general issues related to search reduction techniques is given in [1]. Detailed reviews of the state of the art in the area can be found in [2, 3].

At present, experts widely believe that NP-complete problems are indeed hard to solve. This point of view necessitates a review of the general strategy of research in the area of the algorithmics of NP-hard problems. Attempts to solve NP-complete problems exactly and effectively have been widely replaced by solution search methods based on the investigation of the specific properties of individual NP-complete problems.

Approximate search procedures looking for "a good solution in acceptable time", instead of an exact one, are now in wide use. This approach and the limits of its applicability are discussed in [3]. A detailed review of results on approximate techniques for solving combinatorial NP-complete problems can be found in [4]. An

©1996 American Mathematical Society

extensive literature is devoted to the description of polynomially solvable subclasses of NP-complete problems. A good example of results obtained in this direction is given in [5]. A fairly complete review of subclasses solvable in polynomial time can be found in the reference section of [2].

Still attracting wide attention are methods of local (neighborhood) search based on the classic idea of optimization: the method of trial and error. The primary difficulty in applying techniques of this type to a specific search problem is that of selecting a "reasonable" system of neighborhoods in the space of admissible solutions — this difficulty is closely related to the notion of a "natural" perturbation of an admissible solution. An up-to-date exposition of general ideas as well as some specific applications of local search to well-known optimization search problems can be found in [3].

2. A radically new type of methods of searching for a solution in NP-complete problems (iterative methods) was suggested by Maslov in [6, 7]. These methods can be regarded as an approximation of the strategy of increasing the freedom of choice (IFC-strategy) formulated in [8]. They combine a mechanism for expanding the search area (from integers to rational numbers), the principle of global information processing, and an iterative scheme for reaching the solution. Applied to the classic NP-complete problem of verifying propositional satisfiability for CNF-formulas, the method is formulated as follows.

Consider a CNF-formula F in n variables $a_1, \ldots, a_n$. Variables appearing in F and their negations will be called *literals*. With every literal in F we associate a nonnegative number $\xi(a)$, called the *deferrence* of the literal a. The vector $\xi = \langle \xi(a_1), \ldots, \xi(a_n), \xi(\bar{a}_1), \ldots, \xi(\bar{a}_n) \rangle$ is called the *obstacle*. An obstacle ξ defines a Boolean vector $X = \langle x_1, \ldots, x_n \rangle$ if the following conditions are satisfied:

$$(x_i = 1) \Rightarrow (\xi(a_i) = 0),$$
$$(x_i = 0) \Rightarrow (\xi(\bar{a}_i) = 0), \qquad i = 1, \ldots, n.$$

An obstacle ξ is said to be *correct* if it defines a unique Boolean vector. Let F be of the form $\bigwedge_k D_k$, where the D_k are clauses. An *iterative method* for F is given by an operator $K_{R,L}$ of the type $R_+^{2n} \to R_+^{2n}$ which is defined by $2n$ formulas of the form

$$\xi'(a) = R \cdot \xi(a) + L \cdot \sum_{D_k \ni \bar{a}} \min_{b \in D_k, b \neq \bar{a}} \xi(b).$$

Here a, b are literals from F, and R, L are nonnegative constants.

Evidently, $K_{R,L}$ is a piecewise linear continuous homogeneous operator. In Maslov's works [6, 7] the following statements are formulated:

(1) For any R and L the space of obstacles defining the satisfiable Boolean vector for formula F is an eigenspace of the operator $K_{R,L}$ (for $R \neq 0$ iterations of $K_{R,L}$ preserve correctness of the obstacles defining satisfiable vectors).

(2) For satisfiable Tseitin's formulas in any correct initial approximation the iteration sequence of $K_{R,L}$ converges to a correct obstacle defining satisfiable vectors.

(3) Let F be a satisfiable 2-CNF-formula. For any nonzero R and L and any correct initial approximation the iteration sequence of $K_{R,L}$ converges to an obstacle defining satisfiable vectors.

It follows from (1) that seeking a solution to the satisfiability problem by the iterative method can be reduced to finding linear eigenspaces of $K_{R,L}$. By organizing such a process in different ways and supplementing it with a splitting over most deferred variables, one can obtain a wide class of iterative methods.

3. The articles in this book explore new approaches to the solution of NP-problems. These results were presented at the Seminar "Problems of Search Reduction" held in Leningrad by the "Mathematical Logic in Artificial Intelligence" Commission of the Scientific Council for Cybernetics of the USSR Academy of Sciences. The seminar was originally established to study the heritage of Maslov in the area of iterative methods of search reduction. In 1982–85 the seminar discussed a series of new results on the algorithmics of NP-hard problems.

Let us begin by summarizing the papers containing new results on Maslov's iterative method. In his article, V. Ya. Kreinovich derives the form of Maslov's operator and describes in exact terms the class of such operators. It is proved that the operator $K_{R,L}$ and other its modifications can be constructed with the help of the mechanism of solving the satisfiability problem involving extension of the domain to a "continuous" one. In the article by M. I. Zakharevich it is proved that convergence in the mean always holds for Maslov's operator, thus yielding a new class of fast iterative procedures for seeking the maximal eigenvector. The paper by I. V. Melichev, who experimentally tested the iterative method, discusses a number of problems related to the selection of parameters for Maslov's operator and suggests ways of how to handle these problems.

The article by E. Ya. Dantsin provides an up-to-date review of results on the methods of solving one of the most popular NP-complete problems — the problem of satisfiability. Part of the review, including methods of utilizing the symmetry of formulas, is based on the results of the present author. The article by G. V. Davydov and I. M. Davydova suggests a search reduction technique based on the notion (introduced by the authors) of a dual structure that compactly represents the class of search trees. The optimality criterion suggested in the paper is similar to the criterion based on the duality theorem in linear programming. The article by Yu. V. Matiyasevich proposes a system of nontraditional models for the NP-complete problem of propositional satisfiability. N. K. Kossovsky constructs in his paper an example of a sequence of Boolean functions whose schematic complexity is exponentially bounded from below; the approach is based on the solvability of logical-arithmetical equations with exponential bounds on the solution length.

The articles by A. Ya. Dikovsky and M. I. Kanovich deal with methods of constructing rapid algorithms for solving a number of problems arising in program synthesis in relational databases, program synthesis systems, deductive systems, etc. The paper by A. Ya. Dikovsky gives a classification of computational models by the type of functional dependencies and describes a method for constructing effective algorithms in program analysis and synthesis. M. I. Kanovich introduces in his article the notion of effective calculus, which is then successfully applied to search reduction.

We conclude by mentioning that the papers in this book are of not just theoretical but also applied significance. Individual algorithms proposed in the papers by A. Ya. Dikovsky and M. I. Kanovich are record-breaking in efficiency and can find wide application in practical programming systems. Research on Maslov's iterative method opens up new promising direction in the development of applied systems aimed at solving hard search problems.

References

[1] G. M. Adel'son-Velsky and A. O. Slisenko, *What can we do with problems of exhaustive search?* Algorithms in Modern Mathematics and Computer Science (Proc. al-Khwarizmi Sympos., Urgench, 1979; A. P. Ershov and D. E. Knuth, eds.), Lecture Notes in Computer Sci., vol. 122, Springer-Verlag, Berlin, 1983, pp. 315-342.

[2] M. Garey and D. Johnson, *Computers and intractability: a guide to the theory of NP-completeness*, Freeman, San Francisco, 1979.

[3] Ch. H. Papadimitriou and K. Steiglitz, *Combinatorial optimization: algorithms and complexity*, Prentice-Hall, Englewood Cliffs, NJ, 1982.

[4] A. A. Korbut and Yu. Yu. Finkelstein, *Approximate methods of discrete programming*, Izv. Akad. Nauk SSSR Tekhn. Kibernet. **1983**, no. 1, 165-176; English transl., Engrg. Cybernetics **21** (1983), 124–134.

[5] A. O. Slisenko, *Context-free grammars as a tool for describing polynomial time subclasses of hard problems*, Inform. Processing Lett. **14** (1982), 52-56.

[6] S. Yu. Maslov, *Iterative methods in intractable problems as a model of intuitive methods*, Abstracts 9th All-Union Sympos. Cybernetics, 1981, pp. 52-56. (Russian)

[7] _____, *Asymmetry of cognitive mechanisms and its implications*, Semiotics and Information Science, No. 20, VINITI, Moscow, 1983, pp. 3–31. (Russian)

[8] _____, *Calculi with monotone deductions and their economic interpretation*, Zap. Nauchn. Sem. Leningrad. Otdel. Mat. Inst. Steklov. (LOMI) **88** (1979), 90–105; English transl. in J. Soviet Math. **20** (1982), no. 2.

Algorithmics of Propositional Satisfiability Problems

E. Ya. Dantsin

ABSTRACT. The article considers the problem of propositional satisfiability and surveys methods used for its solution. Classes of formulas for which these methods are effective are described. The relationship with systems of proving unsatisfiability (or tautology) is discussed, and results concerning such systems are given.

Introduction

The satisfiability problem is usually stated as follows: given a propositional formula, one must establish whether it is satisfiable, i.e., true for at least one assignment of variable values, or unsatisfiable. Wider formulations are also considered: for a satisfiable formula to find a satisfying vector; for a unsatisfiable formula to give a proof of unsatisfiability in some formal system.

In theoretical studies, where search problems are more often stated as problems of recognition, one speaks about recognizing both satisfiability and unsatisfiability. Under the deterministic approach to solving the problem they are virtually the same. For each formula the recognition algorithm outputs one of the two answers: "satisfiable" or "unsatisfiable", and in both cases the computation protocol can be regarded as the proof of the validity of the answer. The symmetry between satisfiability and unsatisfiability is broken when one considers nondeterministic solutions.

The only requirement for a procedure that recognizes satisfiability nondeterministically is that for every satisfiable formula a computation must be possible that outputs the answer "satisfiable", and that for all other formulas such a computation must be impossible (proofs of unsatisfiability are not necessarily to be obtained). The most natural procedure of this kind is as follows: given an input formula, select nondeterministically a set of values for its variables and compute the corresponding value of the formula; if the result is "true", output the answer "satisfiable". The running time is bounded by a polynomial in the input formula size, which means that satisfiability is recognized nondeterministically in polynomial time. In the currently accepted terminology the problem of recognizing satisfiability belongs to the class NP whereas the complementary problem of recognizing unsatisfiability belongs to the class $coNP$.

©1996 American Mathematical Society

Two questions remain open:

(1) Is there a polynomial algorithm solving the satisfiability problem, i.e., is it a problem of the class P?

(2) Can one recognize unsatisfiability nondeterministically (the computations must be possible for all unsatisfiable formulas and for them only) in polynomial time? In other words, does the problem of recognizing unsatisfiability belong to the class NP?

The significance of these questions is established by S. Cook's theorem on NP-completeness of the satisfiability problem [1]: if there is an algorithm that recognizes satisfiable formulas then for any other problem in NP one can construct an algorithm of the same time complexity up to a polynomial. The theorem implies that the affirmative answer to question (1) is equivalent to $P = NP$ and equivalent to $P = coNP$, while the affirmative answer to question (2) is equivalent to $NP = coNP$ (thus a negative answer to question (2) implies $P \neq NP$).

The fundamental role of the relationship between the classes P, NP, and $coNP$ is widely known and well covered in the literature (see, for example, [2]). Many NP-complete problems, including the satisfiability problem, have undergone intensive study both for the purpose of clarifying this relationship, and to obtain practically effective algorithms. Even if studies of specific problems did not produce any noticeable progress in finding answers to questions (1) and (2), they produced results useful from the applied viewpoint.

With respect to the satisfiability problem, different approaches were developed identifying "efficiency" and "nonefficiency" areas, and various techniques of proving unsatisfiability (tautology) were studied and their efficiency estimated. The purpose of this article is to discuss these questions and survey the corresponding results. We do not include Maslov's iterative method, which is considered in several other articles in this book.

Wherever possible the author made an effort to stick to accepted terminology. When speaking about algorithms, their complexity, graphs etc, the word "accepted" means the terminology from [3]. For a computational model we use a random access memory machine with logarithmic command weight. In several instances the author was unable to structure his exposition in such a way that the reader could be referred to the literature, and had to define some of the known notions from scratch.

An important contribution to better understanding of the subject were frequent discussions with participants of the Leningrad seminars on Mathematical Logic (held at the Leningrad Branch of the Steklov Mathematical Institute) and on Algorithmics of NP-hard Problems. The author's deep gratitude is due to all of them.

1. Reduction classes and polynomial classes

The satisfiability problem is NP-complete. It is therefore natural to seek solution techniques for separate classes of formulas, trying to achieve effectiveness by utilizing individual properties of formulas belonging to the class. As a rule, by a class we mean a set of formulas defined by a condition expressed in terms of syntactic parameters of the formulas; the condition should be sufficiently easy to verify.

In what follows many examples of different classes will be given, the most famous of them in this section.

For some classes the satisfiability problem can be solved effectively, for others it turns out that it is no easier than for all formulas. Accordingly, two types of classes are defined. A class is said to be *polynomial* if there exists an algorithm recognizing satisfiability of formulas belonging to this class in polynomial time. A class K is said to be a *reduction class* if there exists a polynomial algorithm which for any propositional formula F constructs a formula in K which is satisfiable if and only if F is satisfiable.

1.1. Conjunctive normal forms. Suppose one has to solve the satisfiability problem for arbitrary propositional formulas with connectives $\wedge$ (conjunction), $\vee$ (disjunction), $\neg$ (negation), $\supset$ (implication), $\equiv$ (equivalence). Is it possible to reduce it to the satisfiability problem for formulas of a more special type? The simplest method of achieving such a reduction is given by logical equivalences together with the rule of equivalent substitution [4, Chapter 3, Section 1]. For example, for every formula one can construct an equivalent formula without $\wedge$ and $\vee$ by using equivalences expressing $\wedge$ and $\vee$ through $\neg$ and $\supset$. The construction is performed in polynomial time, and therefore formulas with $\neg$, $\supset$, $\equiv$ constitute a reduction class.

Other reduction classes can be obtained in a similar manner. However this approach, which is based on the rule of equivalent substitution, does not necessarily produce polynomial algorithms. In particular, the standard method of reducing a formula to conjunctive normal form (see the same section in [4]) is nonpolynomial since substitutions that eliminate $\equiv$ and make use of distributivity may cause exponential growth in formula size. Nevertheless, as is well known, formulas expressed in conjunctive normal form constitute a reduction class. The algorithm establishing this fact will be discussed below.

In the context of the satisfiability problem, formulas expressed in conjunctive normal form are usually considered not as words in the language of propositional calculus but up to the order and placement of parentheses inside conjunctive terms and among them. Our approach will be in agreement with this point of view.

A *literal* is a propositional variable or its negation. Literals of the form x, where x is a variable, will be called *positive* and literals of the form $\neg x$ *negative*. Let a be a literal; if a is a variable x, then the literal $\neg x$ is denoted by $\bar{a}$; if a is of the form $\neg x$, then $\bar{a}$ denotes x. We say that literals a and $\bar{a}$ are *complementary*.

A finite set of literals is called a *clause*. An empty clause is denoted by $\square$. By the *length* of a clause we mean its cardinal number. A clause of length 1 is called a 1-clause. A clause containing only positive literals is said to be *positive* and a clause containing only negative literals is said to be *negative*.

A *conjunctive normal form* (abbreviated to cnf) is a nonempty set of clauses. Two cnf are said to be equal if they coincide as families of sets. A clause consisting of the literals $a_1, a_2, \ldots, a_k$ will be written as $a_1 a_2 \ldots a_k$, and the cnf with clauses $D_1, \ldots, D_l$ as $\{D_1, \ldots, D_l\}$.

If we say that a cnf is processed by an algorithm, we always assume that the cnf is represented in the memory of a random-access machine in a similar manner, i.e., by a sequence of cells each containing an integer or a symbol (a variable is represented by a positive integer and the literal complementary to it by the corresponding negative integer). The *size* (or *length*) of a cnf is the number of cells in the sequence.

We say that a set of variable values (or simply a *set of values*) is given if there is a mapping of the set of variables into the set of truth values {0 (true), 1 (false)}. In case the value of the variable x is defined, the literal x assumes the same value, and the literal $\neg x$ the opposite value. The set of values for all variables in a cnf defines the value of the cnf itself: the value of a clause is the maximum of the values of its literals, the value of a cnf is the minimum of values for the clauses it contains. We say that a cnf is *satisfiable* if there exists a set of values for its variables for which it assumes the value 1. If there is no such set of values the cnf is said to be *unsatisfiable*.

The set of all cnf will be denotes by CNF. If the length of a cnf does not exceed 3, we say that it is a 3-cnf and denote the class of such cnf by 3-CNF.

THEOREM. *The 3-CNF class is a linear reduction class, i.e., there exists an algorithm with linear time complexity which for every propositional formula constructs a 3-cnf which is satisfiable if and only if the input formula is satisfiable.*

The theorem is well known and for many belongs to the "folklore". Apparently it appeared for the first time in [5]. An idea of the proof is as follows.

Let F be a propositional formula in a "wide" basis of connectives. We associate with every subformula of F a new variable that is not contained in F. Suppose that subformula S consists of subformulas S_1 and S_2 connected by any of the logical connectives, i.e., S is of the form $S_1 \odot S_2$, and let the formulas S, S_1, and S_2 correspond to variables x, x_1, and x_2. We construct a 3-cnf equivalent to the formulas $x \equiv x_1 \odot x_2$. For example, if S is $S_1 \vee S_2$ we have

$$\{\bar{x}x_1x_2, x\bar{x}_1, x\bar{x}_2\}.$$

All nonatomic subformulas in F are treated in the same way, including subformulas of the form $\neg S$. Take the conjunction of all 3-cnf thus obtained and add the clause consisting of a single variable corresponding to the original formula F. The resulting set is a 3-cnf which is satisfiable if and only if F is. The construction time depends linearly on the size of the input formula.

One can consider smaller reduction class, for example, the class of cnf in which every clause is of length 3, or the class of 3-cnf in which the number of appearances of every variable is limited, etc. (see [2]). A convenient tool for constructing reduction classes is given by a theorem in [6] which will be discussed in Section 1.5 below.

1.2. The class of 2-cnf. One of the best-known polynomial classes consists of 2-cnf, i.e., of those cnf for which the length of every clause does not exceed 2. We shall denote this class by 2-CNF. The polynomial property for 2-CNF (observed, e.g., in [1]) can easily be established in different ways. For example, one can successively eliminate variables by applying the resolution rule, which for 2-cnf yields only clauses whose length does not exceed 2. The author has not been able to establish whether a linear algorithm was ever published. We suggest the following proof of linearity for 2-CNF based on the algorithm identifying connected components.

THEOREM. *There exists an algorithm which recognizes satisfiable 2-cnf in linear time and constructs satisfying sets for them.*

PROOF. Let E be a 2-cnf. We can consider it as a set of implications: a 1-clause a corresponds to the implication $\bar{a} \supset a$, a clause of the form ab corresponds to the implications $\bar{a} \supset b$ and $\bar{b} \supset a$. We associate with E a labeled oriented graph G whose nodes are all literals in E and literals complementary to them and whose edges are defined by implications: an implication $a \supset b$ defines an oriented edge (a, b).

We shall prove that E is satisfiable if and only if there is no cycle in G passing through a complementary pair of literals. Consider a set of values for variables in E and assign to each node of the graph the value of the corresponding literal, i.e., either 0 or 1. Suppose there is a complementary pair of literals connected by a cycle. Then under any assignment this cycle has an edge starting from 1 and ending at 0. Therefore, the implication corresponding to this edge is false and E is unsatisfiable.

Assuming that there are no cycles with complementary pairs, we now show how a satisfying set can be constructed. Two nodes in a oriented graph are said to be equivalent if there is a cycle passing through them. The corresponding equivalence classes are called strongly connected components. If two complementary nodes belong to different components G_1 and G_2, then these components are dual: nodes of G_1 are complementary to nodes of G_2, while every edge (a, b) in one component corresponds to the edge $(\bar{b}, \bar{a})$ in the other. Since we assume that no component contains complementary literals, the set of components falls into pairs of dual components. Let G_1 and G_2 be such a pair. These components are either not connected by a path, or there is a path leading from one component into another but no path back. In case there is no path at all, select either of the components G_1 or G_2 and assign 0 to all its nodes and 1 to the complementary literals in the other component. In case a connecting path exists, assign 0 to the component it starts from and 1 to the component it ends at. The values so assigned define a satisfying set since there are no edges starting from 1 and ending at 0, and so all implications are true.

Thus in order to recognize satisfiability it is sufficient to: (1) construct the graph corresponding to the 2-cnf; (2) construct its strongly connected components; (3) check whether there is a component with complementary literals. If such a component exists, output the answer "unsatisfiable"; if not, output the answer "satisfiable". These operations are performed in linear time; in particular, strongly connected components are constructed by the linear algorithm 5.4 in [3].

1.3. Horn's cnf. We define two classes of cnf. The class H^- consists of cnf for which every clause has at most one positive literal. The class H^+ consists of cnf for which every clause contains at most one negative literal. A cnf is said to be a *Horn's cnf* if it belongs to either H^- or H^+. For a Horn's cnf it is known that the satisfiability problem can be solved in linear time (e.g., [7]). Here is an algorithm which is not the fastest (asymptotically) but is very easy to describe.

THEOREM. *There is an algorithm that recognizes a Horn's cnf in polynomial time and constructs a satisfying set.*

PROOF. The main part of the algorithm is the following recursive procedure which takes a Horn's cnf as its input.

(1) Select a 1-clause. If the Horn's cnf contains no 1-clauses, it is satisfiable: for a cnf belonging to H^- all clauses are true for 0 values of variables; for a cnf in H^+, for values of all variables equal to 1.

(2) Let a be the selected 1-clause. Remove all the clauses containing the literal a from the cnf (if a is contained in all clauses, the initial cnf is satisfiable).

(3) Remove the literal $\bar{a}$ from the remaining clauses (if this results in an empty clause, the initial cnf is unsatisfiable).

(4) Apply the above procedure to the cnf thus obtained (it is a Horn's cnf of the same class H^- or H^+ as the initial one).

The satisfying set can be constructed simultaneously by associating with every clause the variable value: whenever a positive 1-clause is selected, the corresponding variable is given the value 1; for negative 1-clauses, the value 0.

The number of steps in the procedure does not exceed the number of variables in the initial cnf.

1.4. Systems of linear equations. Another polynomial class, denoted by $\mathcal{E}$, consists of cnf representing systems of linear equations over the two-element field $\{0, 1\}$ (see [**8, 9**]). Let $x_1, \ldots, x_k$ be variables, let ε denote 0 or 1, and let

(1.4.1) $$x_1 + \ldots + x_k = \varepsilon$$

be the equation over the field $\{0, 1\}$. Consider a cnf E with variables $x_1, \ldots, x_k$ consisting of 2^{k-1} clauses in every one of which:

(1) there are exactly k literals, and every variable x_i appears in the clause once and only once;

(2) the number of negative clauses is not equal to ε modulo 2.

We say that the cnf E *represents* the equation (1.4.1). For example, the equation

$$x_1 + x_2 + x_3 = 0$$

is represented by the cnf

$$\{\bar{x}_1 x_2 x_3, x_1 \bar{x}_2 x_3, x_1 x_2 \bar{x}_3, \bar{x}_1 \bar{x}_2 \bar{x}_3\}.$$

It is not hard to see that the set of solutions for equation (1.4.1) coincides with the collection of sets satisfying E.

Given a system of linear equations over $\{0, 1\}$, the union of cnf representing these equations forms a cnf which *represents* the system. A set of variable values of the variables solves the system if and only if it satisfies the representing cnf. The set of cnf representing systems of linear equations will be denoted by $\mathcal{E}$.

THEOREM. *There exists an algorithm that recognizes satisfiability of a cnf from class $\mathcal{E}$ in polynomial time and constructs a satisfying set.*

The proof follows from the fact that systems of linear equations can be solved in polynomial time ([**10**]). The algorithm can be based on the Gauss method for solving linear equations. A different technique for recognizing satisfiability of cnf from $\mathcal{E}$ is described in [**9**].

1.5. Classes generated by bases. In practical applications one usually has to solve the satisfiability problem not for arbitrary formulas but for specific classes. Before seeking an efficient algorithm for a given class (or after failing to find one), it makes sense to investigate whether the class is a reduction class. In this case a very useful tool may be provided by the theorem from [6] which gives a simple criterion for whether a set of formulas is a reduction class. We shall state the theorem in a slightly different but similar version (see also [2, Section A9.1]).

Let X and Y be two sets of variables, and r a one-to-one mapping of X onto Y. Let E be a cnf with the set of variables X. Replace every positive literal $x \in X$ in E by the literal rx and every negative literal $\bar{x}$ by $r\bar{x}$. The resulting formula will be said to have been obtained from E by *renaming* the variables r.

Let B be a finite set of cnf which we shall call a *basis*. We say that a cnf E is *generated* by the basis B if E is the union of cnf $E_1, \ldots, E_k$, where each E_i is obtained from some cnf by renaming variables. Such a cnf E is considered defined if for every cnf in the basis a set (maybe empty) of renamings $r_1, \ldots, r_l$ is given yielding the "summands" $E_{i_1}, \ldots, E_{i_l}$ of the union. We note that the class of cnf generated by a fixed basis contains arbitrarily long cnf.

The statement of the theorem makes use of classes discussed above (the class 2-CNF, Horn's classes H^- and H^+, the class $\mathcal{E}$ representing systems of linear equations) as well as two new classes. The first new class consists of cnf which we shall describe as satisfied by the 0-*set*, i.e., by the set in which all variables assume the value 0. A cnf belongs to this class if and only if each of its clauses contains at least one negative literal. The second class is the class of cnf satisfied by the 1-*set* in which the value of every variable is 1. Every clause in such a cnf contains at least one positive literal.

THEOREM. *Let B be a finite set of cnf (a basis) and S_B the class of cnf generated by this basis. If B satisfies at least one of the six conditions listed below, then it is a polynomial class, i.e., the satisfiability problem for S_B can be solved in polynomial time. Otherwise S_B is a reduction class.*

1. $B \subset 2\text{-}CNF$;
2. $B \subset H^-$;
3. $B \subset H^+$;
4. $B \subset \mathcal{E}$;
5. $B \subset \{cnf\ satisfied\ by\ the\ 0\text{-}set\}$;
6. $B \subset \{cnf\ satisfied\ by\ the\ 1\text{-}set\}$.

The proof, based on Post's theorem on closed classes of the algebra of logic, is given in [6].

2. Splitting method

Some of the most popular techniques for solving the satisfiability problem are based on the splitting method, or the method of resolutions. The first procedure together with the splitting rule was proposed by Davis and Putnam in 1960 [11]. Efficiency of the splitting method was studied by Z. Galil in [3] and by the present author in [13, 14]. The method of resolutions suggested by J. Robinson in 1965 was investigated much more actively. An ample literature is devoted to this approach (dealing, in truth, mostly with applications to predicate logic); an extensive list of references is given in [16].

It is partly for this reason that we devote more attention to the splitting method (the resolution approach and its relation to splittings will be treated in Section 3).

Another reason is the important difference between the two methods. Namely, seeking a deduction procedure in the calculus of resolutions is aimed at proving unsatisfiability, while splitting procedures output either a satisfying set or a proof that no such set exists. The splitting method can therefore also be used in applications where the formulas under consideration are known to be satisfiable and it only remains to find a satisfying set.

2.1. Splitting tree and tactics for its construction. Let us agree to represent a collection of variable values by a set of literals true on this collection: if a variable is assigned the value 0, then it is represented by the literal $\bar{x}$; if it is assigned the value 1, it is represented by the literal x. Thus a collection is a set of literals containing no complementary pairs. We shall write such a literal in the same manner as a clause, i.e., as a chain of literals $a_1, \ldots, a_k$.

The collection of values of all variables for a cnf defines the value of the cnf itself (see Section 1.1 above). Let us extend the notion of a value for a cnf to include the case when values are not necessarily assigned to all variables. Consider a clause D and a collection S. If D contains at least one literal from S, then D is considered to be true, i.e., to assume the value 1. Otherwise the value of D on S is defined as the clause obtained from D by dropping literals that are false on S (for every such literal the complementary one is also contained in S). In particular, the value may turn out to be an empty clause $\Box$; in this case it is said that D is false on S. The value of a cnf E on a collection S is denoted by $E[S]$. In order to compute $E[S]$ one has to compute the values of the clauses of E on S: (1) if these values include $\Box$ then E is false on S, i.e., $E[S]$ is 0; (2) if the values of all clauses are equal to 1 then E is true on S, i.e., $E[S]$ is 1; (3) if $E[S] \neq 0$ and $E[S] \neq 1$, then $E[S]$ is a cnf formed by the values of clauses other than 1. (This definition extends the notion of a satisfying set for E; it is a collection S such that $E[S] = 1$).

By a *splitting* of a cnf along a literal a we mean the computation of the values of E on 1-collections a and $\bar{a}$. By the result of the spitting we mean $E[a]$ and $E[\bar{a}]$. The variable corresponding to the literal a is said to be the splitting variable. The *splitting tree* for a cnf E is a labelled binary tree in which

1. the cnf E is assigned to the root;
2. if the cnf E_v is assigned to the node v, then the descendants of v are assigned the results of splitting E_v along one of its variables x: the left descendant is assigned $E_v[x]$, and the right one is assigned $E_v[\bar{x}]$; the corresponding edges are assigned x and $\bar{x}$ respectively;
3. the leaves are assigned truth values 0 and 1.

We shall also assume that to each node v there corresponds a collection S_v which contains those and only those literals that are assigned to the edges forming a path from the root to the node v. The definition implies that the label of the node v is $E[S_v]$.

The significance of the notion of the splitting tree for the satisfiability problem is illustrated by the following statements, which can be proved easily. Let T be an arbitrary splitting tree for E.

1. To every leaf with label 1 there corresponds a collection satisfying E.
2. For every collection S satisfying E and including the values of all its variables there is a leaf with label 1 in T corresponding to a collection S' such that $S' \subseteq S$.

3. If 0 is assigned to all leaves of T, then E is unsatisfiable.

4. If E is unsatisfiable, then in any splitting tree for E all leaves have label 0.

In order to construct the splitting tree it is sufficient to define an algorithm which selects a splitting variable at every node. Such an algorithm will be referred to as the *tactics* for constructing the splitting tree. Examples of tactics: splitting along a variable contained in a clause of the least length; splitting along a variable having the largest number of occurrences in a given cnf; splitting along a variable having the greatest absolute value of the difference between the number of positive literals and that of negative literals, and so on.

Consider the first example. The heuristics of the situation suggests that this tactics makes sense for cnf containing relatively many clauses of lengths 1 and 2, especially if one takes into account the fact that every splitting increases the number of such clauses. The polynomial algorithm for a Horn's cnf (Section 1.3) actually reduces to the splitting tactics along variables from 1-clauses. For a 2-cnf one can also construct a polynomial algorithm which creates 1-clauses after each splitting (such a sequence of splittings corresponds to the path in the graph described in Section 1.2).

Let T be a splitting tree for a 3-cnf in which every splitting is done along a variable of the least length. The upper bound on the number of nodes in T is of course exponential but somewhat lower than the obvious estimate 2^{n+1} (where n is the number of variables in the cnf); namely, one can prove that the number of nodes in T is less than λ^{n+3}, where $\lambda = 1.618\ldots$ (the "golden section").

A tactic that can be applied to a 3-cnf performs the splitting along a variable which occurs most often in clauses of length 3. The splitting result in a number of 2-cnf and their satisfiability is established by some known algorithm (e.g., the one given in Section 1.2). This tactic, as well as many others, should be combined with eliminating 1-clauses, i.e., one has to split the cnf along the variables contained in 1-clauses (if any). It also makes sense to "process" cnf obtained as a result of the splitting procedure. "Processing" means transformations which in some sense simplify the cnf but do not affect its satisfiability properties. We list a few such transformations:

1. eliminating clauses containing another clause as a subset;
2. eliminating nonessential clauses (a clause D is said to be *essential* if for any literal a in it there is a clause $\bar{a}B$ such that the union $D \cup B$ contains no complementary pairs, see [9]);
3. contracting resolution, i.e., replacing clauses of the form aB and $\bar{a}C$, where $B \subseteq C$, by the clause C, etc.;

the list of such transformations can easily be continued.

It follows from the results published in [12] and discussed below that any algorithm sharply constructing a splitting tree has exponential complexity on some class of cnf. Using the same ideas it is apparently not hard to find examples realizing the exponential case also for modified versions of these algorithms including processing. Nevertheless, procedures constructing a tree with splittings and performing processing at the nodes are easily programmed and seem to be quite suitable for practical applications.

2.2. The graph of variables. Whenever a splitting tree is being constructed, one can expect that equal cnf may appear at different nodes. Clearly, to check satisfiability it is sufficient to "split to leaves" only one such node. The procedure of constructing the tree can be modified in such a way that no cnf is split twice. This can be achieved by constructing the set of cnf created in the course of splitting, and then checking whether any new cnf is already contained in this set. The checking procedure will be fast if this set is organized in the form of a *dictionary* (see [**3**, Chapter 4]), i.e., as a linearly ordered set with the structure of a balanced search tree (in a dictionary containing m entries any search or add operation requires no more than $O(\log m)$ comparisons of elements).

In [**13**] we investigate those properties of cnf which make it effective to apply procedures based on splittings taking equality into account. These properties can be expressed quantitatively and make it possible to estimate the running time for the procedure, the estimate being obtained very fast from the cnf itself. Let us state these results.

We say that two variables in a cnf are *adjacent* if both of them appear in any of its clauses (with any signs). The *graph of the variables* of a cnf E is a labeled nonoriented graph whose set of nodes coincides with the set of variables of E and any two of whose nodes are connected by an edge if the corresponding variables are adjacent in E. We shall assume that if a cnf contains n variables, then they are numbered by integers 1 to n (in a RAM the variables are represented by their numbers). Consider the graph of variables G for a cnf with n variables. By G_i, where $1 \leq i \leq n$, we denote the subgraph of G generated by variables whose numbers do not exceed i. Denote by δ_i the number of nodes on the boundary of G_i (the *boundary* of a subgraph is the subset of its nodes such that every one of them is adjacent to at least one node not contained in the subgraph). The numbers $\delta_1, \ldots, \delta_n$ will be called *boundary characteristics* of the cnf (we remind the reader that the numbering of variables is fixed).

THEOREM. *The boundary characteristics are computed from a cnf in linear time.*

THEOREM. *There is an algorithm (based on splittings and taking into account equality of cnf) which recognizes satisfiable cnf and constructs satisfiable sets in time $O(Nn^2 2^\delta)$, where N is the cnf size, n the number of variables, and δ the maximum of its boundary characteristics.*

Two algorithms—one computing boundary characteristics and the other other recognizing satisfiability—are given in [**13**]. The latter is in fact the above modification of the procedure for constructing the splitting tree. The construction tactics prescription is to select for each splitting the variable having the least number.

It follows from the second theorem that every positive constant c generates a polynomial class consisting of cnf whose boundary characteristics do not exceed $c \log_2 N$, where N is the cnf size. In particular, such are classes of cnf with *locally contained* variables. This name refers to any class of cnf whose clauses are ordered in such a way that the differences of numbers corresponding to clauses containing the same variable are bounded by a constant. A fast algorithm recognizing satisfiability for cnf with locally contained variables was suggested in [**17**].

The estimate given by the second theorem is trivial when the boundary characteristics of the cnf are close to the number of its variables. Nevertheless, even in this case individual properties of the graph of variables (e.g., the possibility of splitting it into subgraphs that are weakly "chained" to each other) may indicate an appropriate effective tactic for the splitting method ([14]).

2.3. The utilization of symmetry. Consider a one-to-one mapping of the set of all literals onto itself which has the following "parity" property: if a literal a goes into the literal b, then $\bar{a}$ goes into $\bar{b}$. Such mappings will be called *isomorphic transformations*. Their action on collections of literals and cnf is defined in a natural way. Two cnf E_1 and E_2 are said to be *isomorphic* if there is an isomorphic transformation that takes E_1 into E_2. For example, the following three cnf are isomorphic:

$$\{xy, \bar{x}y, \bar{y}\}, \quad \{\bar{x}\bar{y}, \bar{x}y, x\}, \quad \{yz, \bar{y}z, \bar{z}\}.$$

Clearly, isomorphic transformations preserve the property of satisfiability.

An *automorphism* of a cnf is an isomorphic transformation taking the cnf into itself. The automorphisms of a cnf E form a group denoted by $\mathrm{Aut}E$.

The group $\mathrm{Aut}E$ acts on the set of collections of variable values of E. The set of collections is stratified into orbits: two collections belong to the same orbit if there is an element in $\mathrm{Aut}E$ taking one collection into the other. The utilization of symmetry (which is represented by automorphisms) in the satisfiability problem is based on the following assertion: when substituted in E, collections from the same orbit produce isomorphic cnf, i.e., if collections S_1 and S_2 belong to the same orbit, then $E[S_1]$ is isomorphic to $E[S_2]$.

The procedure for using the symmetry could be as follows. Let E be a cnf with n variables. The action of the group $\mathrm{Aut}E$ splits the set consisting of the 2^n collections of values of all variables in E into l orbits. Select collections $S_1, \ldots, S_l$ representing them and compute the values of E on them. The cnf E is satisfiable if and only if at least one of the resulting values is "true". However, even if l is substantially smaller than 2^n this procedure does not necessarily produce an effective algorithm, since the computational complexity of the problem of selecting representatives for each orbit may be higher than that of the initial problem itself.

In the splitting method the symmetry is used in a somewhat different manner. To keep it simple, one can say the more automorphisms for the initial cnf, the more isomorphic cnf arise in the process of constructing its splitting tree. Isomorphic cnf can be "pasted" together as equal cnf were "pasted" together in Section 2.2, and then only one of isomorphic cnf split "to the leaves". However, the method that takes equality into account cannot be directly transferred to the case of isomorphism, since cnf equality can be recognized really fast, while checking cnf for isomorphism is a problem whose complexity is equivalent to checking graph isomorphism. To be able to "paste" isomorphic cnf in an organized manner one has to know automorphisms of the initial cnf and how they are affected by splittings. Let us illustrate the situation by a few examples:

(1) Suppose that a cnf E includes a variable x which goes into $\bar{x}$ under the action of an automorphism from $\mathrm{Aut}E$, i.e., that literals x and $\bar{x}$ belong to the same orbit. Then the results of splitting E along x and $\bar{x}$ are isomorphic and can be pasted together. The cnf actually reduces to another one which has one less variable.

(2) Suppose that the action of an automorphism from $\mathrm{Aut}E$ takes literal a into literal b and at the same time takes literal b into a. Then the collections $a\bar{b}$ and $\bar{a}b$ belong to the same orbit, and therefore if one is splitting E along a and b the

nodes labeled $E[a\bar{b}]$ and $E[\bar{a}b]$ can be pasted together. This yields a reduction to not four but three cnf: $E[ab]$, $E[\bar{a}b]$, and $E[\bar{a}\bar{b}]$.

(3) A similar situation arises when E has an isomorphism acting on the literals a, b, and c as a cyclic permutation (a, b, c). It is not hard to see that when splitting E along these literals the nodes corresponding to the collections $\bar{a}bc$, $a\bar{b}c$, $ab\bar{c}$ are merged, and so are the nodes corresponding to the collections $\bar{a}\bar{b}c$, $\bar{a}b\bar{c}$, $a\bar{b}\bar{c}$. Therefore, out of the eight cnf arising in the splittings, it is sufficient to check satisfiability of the following four:

$$E[abc], E[\bar{a}bc], E[\bar{a}\bar{b}c], E[\bar{a}\bar{b}\bar{c}],$$

each of which has at least three less variables.

Consider a generalization of the last two examples. Let $\{x_1, \ldots, x_m\}$ be a subset of variables of the cnf E. By a splitting of E along this subset we mean the computation of the values of E on all 2^m collections of length m of the variables $x_1, \ldots, x_m$. The set of these values may include isomorphic ones. Suppose that after factoring them out we obtain exactly l pairwise nonisomorphic representatives $E_1, \ldots, E_l$. In this case we say that $E_1, \ldots, E_l$ result from a splitting *up to an isomorphism*.

The subset $\{x_1, \ldots, x_m\}$ will be called *symmetric* if the group of automorphisms Aut$E$ acts on it as the symmetric group of order m—in other words, if for any subsequence $x_{i_1}, \ldots, x_{i_m}$ there exists an automorphism taking x_k into x_{i_k} for all k from 1 to m. We note that in order to check whether a given subset is symmetric it is sufficient to verify whether the automorphisms generate some system of generators of the symmetric group of permutations, for example, the cyclic permutations (x_1, x_2) and $(x_1, \ldots, x_m)$.

THEOREM. *Let $\{x_1, \ldots, x_m\}$ be a symmetric subset of variables of the cnf E. Any splitting along this subset up to an isomorphism yields at most $m + 1$ values. These values can be represented by $E[S_0], \ldots, E[S_m]$, where S_i denotes the collection $\bar{x}_1, \ldots, \bar{x}_i, x_{i+1}, \ldots, x_m$.*

The proof follows from the fact that the action of the symmetric group on $\{x_1, \ldots, x_m\}$ splits the set of 2^m collections of length m into $m + 1$ orbits. The collections $S_0, \ldots, S_m$ represent these orbits.

In fact this theorem formulates the splitting rule along a symmetric subset of variables. The greater the cardinal number of the subset, the better the effect of using the splitting along along this subset. One class where such a method is effective consists of cnf expressing in propositional calculus the so-called Dirichlet principle. This class is often cited in literature on satisfiability as an example of a class which is very difficult to handle by either the method of resolutions or the splitting method that does not take isomorphism into account (these issues will be discussed below). An application to these cnf of the splitting rule along a symmetric subset reduces the proposition to a similar proposition of smaller dimension. Such an ingenious version of induction results in polynomial complexity; see [14].

Similar splitting rules can be formulated for some other groups of permutations arising from the action of automorphisms. A rule is effective if the group splits the set consisting of collections of values of all variables into noticeably fewer orbits (on the computation of the number of orbits, see [18]).

3. Proof system

An algorithm solving the satisfiability problem outputs for every input formula one of the two answers: "satisfiable" or "unsatisfiable". The computation protocol can be considered as a proof of the validity of the answer (the precise meaning of this statement will be defined below). The execution time of the algorithm is therefore not less than the time needed to write out a formal proof of satisfiability or unsatisfiability. This explains the interest in proofs and their lengths.

Another reason for this interest is related to the following widely known approach. Consider a deductive system S satisfying at least one of the following requirements:

(1) in S, proofs of satisfiability can be formalized for all satisfiable formulas;

(2) in S, proofs of unsatisfiability can be formalized for all unsatisfiable formulas.

Choose a proof search technique in S having the following property of "completeness": if for the tested formula there are proofs in S, then at least one of them will be found. The absence of the proof is possible only if S satisfies just one of the requirements (1) and (2). If (1) is the case, then the formula is unsatisfiable, and in case (2) it is satisfiable (proofs corresponding to either case can be formalized in some extension of the system S). In fact, both the system S and the search technique in it define an algorithm solving the satisfiability problem. Its effectiveness depends on the "cardinality" of S (the higher the cardinality, the shorter the proofs that are possible in the system), as well as on the length of search at each step; see [19].

The approach just described is well studied for those systems, where proofs are deductions in some calculus (e.g., in the calculus of resolutions; many other examples will be given below). In the literature, it is usually formulated not for the satisfiability problem, but for the dual problem, where one has to check whether the formula is a tautology. This is apparently due to the fact that tautology deductions are more widespread in logical calculi. In our context we can ignore the difference between tautology and unsatisfiability—the result can be stated in parallel in both languages. We shall stick to the unsatisfiability version and "translate" all the results under discussion into this language.

Let X and Y be two alphabets and denote by X^* and Y^* the corresponding sets of words. Let $L \subseteq Y^*$. Following [20] we say that a *proof system* is given for L if there is a mapping from X^* onto L that can be computed by a polynomial algorithm. A word x is said to be a *proof* for y if $f(x) = y$. A proof system (abbreviated to p.s.) is said to be *polynomially bounded* if there exists a polynomial p such that for any word y in L there is a proof x whose length does not exceed $p(|y|)$, where $|y|$ is the length of y.

Polynomially bounded p.s. are known to exist for the set of satisfiable formulas. For example, such a system is given by any algorithm computing the formula values from the variable values in polynomial time: for a satisfiable formula y a proof is the protocol x corresponding to the computation of the value of y on the satisfying set. There also are calculi where all satisfiable formulas (and them only) are deducible and there always exists a deduction whose length is bounded by a polynomial of the formula being deduced ([21]). However, we do not know whether algorithms solving the satisfiability problem by searching for a proof in such systems have ever been studied.

The situation with unsatisfiability is different: the question whether there is a polynomially bounded proof system for unsatisfiable formulas remains open. It is equivalent to question (2) discussed in the introduction. If the answer is affirmative, then $NP = coNP$, while the negative answer implies $P \neq NP$. At the same time many calculi used for deducing unsatisfiable formulas are well studied, and approaches based on proof search in such calculi were and still are very popular. Efficiency estimates of p.s. for unsatisfiable formulas are of both theoretical and practical interest. So are deductive resources enhancing this efficiency.

3.1. Lower bounds. For some specific proof systems for unsatisfiability it is known that they are not polynomially bounded. The first interesting result of this kind was obtained by Tseitin in [8] for the p.s. based on regular resolutions. Before stating it let us make a terminological remark. Here and below, when speaking about a deduction is a calculus we always mean a deduction written in linear form, i.e., in the form of a sequence. The length of the deduction is the number of members in this sequence.

By a *resolution* proof of unsatisfiability for a cnf E we mean a deduction of an empty clause in the calculus of resolutions from clauses in E. Such a proof is said to be *regular* if the corresponding deduction tree satisfies the following condition: there is no path from the root to a leaf on which some variable is resolved more than once. A deduction that is not regular can be transformed into a regular deduction having the same input clauses ([8, Section 1]).

THEOREM. *There exist infinitely many unsatisfiable cnf for each of which the length of any regular resolution proof is greater than $2^{c\sqrt{N}}$, where N is the size of the cnf and c is a constant ($c > \frac{1}{7}$).*

The set of cnf in question is contained in the polynomial class $\mathcal{E}$ (see Section 1.4). The proof of the theorem is outlined in [8], and given in more detail in [22], which uses the same technique but because of the different selection of the subclass of $\mathcal{E}$ raises the bound to 2^{cN}.

A generalization of this theorem to arbitrary resolution proofs follows from [23], where it is proved that if an unsatisfiable cnf admits a regular resolution proof of length l, then it also admits a regular resolution proof of length not greater than l^3. Therefore, proof systems based on the resolution rule only are not polynomially bounded.

Similarly, so are proof systems for which the proof of unsatisfiability is constructed by splittings (without taking into account the cnf isomorphism under splittings). Such proofs can be formalized as deductions in the calculus Σ_0 (its extensions Σ_1 and Σ_2 will be introduced below) which have one axiom and one rule of inference. The axiom is the logical "constant" 0 ("false") while the deduction rule is a so-called splitting rule which can be formally stated as follows: the cnf E is deducible from E_1 and E_2 if both E_1 and E_2 are splitting results for E, i.e., if $E_1 = E[x]$ and $E_2 = E[\bar{x}]$ for some variable x. In Σ_0 all unsatisfiable cnf, and only they, are deducible.

THEOREM. *There exist infinitely many unsatisfiable cnf such that for each of them the length of any deduction in Σ_0 is greater than $2^c N$, where N is the cnf size and c is the same positive constant for all cnf.*

The theorem is proved in [**12**]; the technique developed in [**8**] for regular resolutions is extended to splittings. The set of cnf is the same subclass of $\mathcal{E}$ for which the same bound on the length of a resolution proof is established.

Formalizations of the propositional calculus without the section rule also are not polynomially bounded. This is established in [**24**] for the sequential calculus S_0 whose formulas are constructed from the variables and the symbol $\perp$ for "false" by means of a single connective, implication. The axioms of S_0 are the following sequents:

$$\Gamma, \perp \Rightarrow \Delta,$$
$$F, \Gamma \Rightarrow F, \Delta,$$

where F is a formula, and Γ and Δ are lists of formulas. The rules of inference consist of the rule of introduction of implication in the antecedent and the rule of introduction of implication in the succedent ([**25**]).

THEOREM. *There is no polynomial p such that for any true formula F of size p in S_0 there is a deduction of the sequent $\Rightarrow F$ having length less than $p(n)$.*

3.2. Admissible rules. If we know that a proof system is not polynomially bounded, then in order to make proof lengths shorter it is natural to try to make it more efficient—at least in the case of known examples with exponential lower bound. A widely used tool for shortening proofs in calculi makes use of admissible rules, i.e., inference rules which, when added to the calculus, do not change the set of deducible words. Two classic examples of admissible rules for the propositional calculus are well known in logic: the *section rule* (in various forms) and the *substitution rule*. Sometimes an admissible rule is understood in a wider sense and includes rules whose formulation require changes in the language of the calculus and possibly even in the notion of deduction. A good example is the *extension rule* proposed by Tseitin in [**8**] for the calculus of resolutions.

This rule is a scheme for generating new clauses which can then be used as premises for the rule of resolution. Let S denote the set of clauses already generated in the deduction. An application of the extension rule makes it possible to add to S three arbitrary clauses of the form

$$xa, xb, \bar{x}\bar{a}\bar{b},$$

where it is required that the literals x and $\bar{x}$ are not included in any of the clauses in S. The "extended" set of clauses is satisfiable if and only if S is.

In fact the extension rule provides a useful tool for proving new notions and notation. Applying resolutions to "new" variables is similar to taking sections along them. For the propositional calculus the extension rule means the rule which allows one to add to the deduction formulas of the form $x \equiv A$, where A is an arbitrary formula and x is a variable which does not appear in the preceding part of the deduction [**20**]. This is apparently the most clear-cut formulation.

For the calculus Σ_0 with the splitting rule (see Section 3.1) two admissible rules were suggested: the rule of *thinning*, a one-premise deduction rule in which the antecedent cnf is a subset of the succedent cnf; and the rule of *isomorphic substitution*, also a one-premise rule which allows one to deduce from a cnf any cnf isomorphic to it [14]. The calculus Σ_1 is obtained by adding the rule of thinning to Σ_0, while the calculus Σ_2 is Σ_1 plus the rule of isomorphic substitution.

The rule of thinning makes Σ_1 more convenient than Σ_0 in those cases where one has to deduce several cnf having the same unsatisfiable subset of clauses. Σ_1 is at least as efficient as the proof system based on regular resolutions (without the extension rule). This follows from [12], where it is shown that for any unsatisfiable cnf E there is a deduction in Σ_1 whose length is less than that of any regular resolution proof for E. Nevertheless, the rule of thinning is not efficient enough: the theorem of the exponential lower bound for the length of a proof in Σ_0 is extended to Σ_1 ([12]). The rule of isomorphic substitution allows one to use in deductions in Σ_2 the new symmetry of cnf. It can also be regarded as an analogue of a "limited" substitution rule which allows only literals to be substituted.

3.3. Comparison of efficiency.
In this section we use admissible rules to give results on shortening proof lengths and compare the efficiency of the resulting proof systems. By the *proof length* in a proof system f we mean the function L_f defined by the formula
$$L_f(n) = \max_{|y|=n} \min_{f(x)=y} |x|,$$
where x is a proof for y, and $|x|$ and $|y|$ denote the lengths of the corresponding words. (Polynomial boundedness for a proof system means that proof lengths in it are bounded by a polynomial.)

THEOREM ([8]). *The proof length in the proof system based on resolutions without the extension rule is not bounded by any polynomial in the proof length in a resolution proof system with the extension rule.*

In the proof of this statement it is shown in [8] that for cnf on which the exponential bound for regular deductions is reached there are short deductions with the extension rule. The regularity condition can be dropped ([23]).

THEOREM ([14]). *The proof length in Σ_1 is not bounded by any polynomial in the proof length in Σ_2.*

The rule of isomorphic substitution makes it possible to utilize the symmetry of cnf from class E to obtain shorter deductions for cnf on which the exponential bound of the proof length in Σ_1 is reached.

THEOREM ([24]). *The proof length in the sequential calculus S_0 is not bounded by any polynomial in the proof length in S_0 with the section rule.*

THEOREM([24]). *The proof length in Hilbert's formalization of the propositional calculus with axiom schemes and section rule is not bounded by any polynomial in the proof length in the same calculus with the substitution rule.*

Unlike the three preceding theorems, the proof of the last one yields no example of a formula on which the proof length grows exponentially.

The article [20] contains a review of results on proof systems based on various formalizations of the propositional calculus. We briefly mention some of these results. Let us say that a proof system f_1 *polynomially models* a proof system f_2

if the sets of provable words in f_1 and f_2 coincide and there exists a polynomial algorithm which for every proof in f_2 constructs a proof of the same word in f_1.
1. All Hilbert systems, sequential systems with section, and systems of natural deduction polynomially model each other.
2. If a Hilbert system with extension rule (in the propositional version) is polynomially bounded, then all Hilbert systems with substitution rule are polynomially bounded.
3. If a Hilbert system with substitution rule is polynomially bounded, then all Hilbert systems with substitution rule are polynomially bounded.
4. A Hilbert system with substitution rule polynomially models the same system with extension rule.

We also note that the proof system Σ_1 polynomially models the resolution proof system without extension rule ([12]), and the Hilbert system with extension rule polynomially models the proof system Σ_2 ([27]).

No examples are known demonstrating that Σ_2 and systems with extension rule are not polynomially bounded. Neither is it known whether the proof length in Σ_2 or in systems with extension is bounded by a polynomial in the proof length in systems with substitution. It would be very interesting to compare proof length in these systems with proof lengths in more efficient proof systems, for example, those like arithmetics.

References

1. S. A. Cook, *The complexity of theorem-proving procedure*, Proc. 3rd Annual ACM Symposium on the Theory of Computing, Shaker Heights, Ohio, 1971, pp. 151-159.
2. M. Garey and D. Johnson, *Computers and intractability: a guide to the theory of NP-completeness*, Freeman, San Francisco,CA, 1979.
3. A. Aho, J. Hopcroft, and J. Ullman, *The design and analysis of computer algorithms*, Addison-Wesley, Reading, MA, 1976.
4. D. Hilbert and P. Bernays, *Grundlagen der Mathematik*. I, rev. ed., Springer-Verlag, Berlin, 1968.
5. M. Waisberg, *Untersuchungen über den Aussagenkalkül von A. Heyting*, Wiadom. Mat. **46** (1938), 45–101.
6. T. J. Schafer, *The complexity of satisfiability problems*, Proc. 10th Annual ACM Sympos. Theory of Computing, ACM, New York, 1978, pp. 216-226.
7. M. I. Kanovich, *Effective calculi as a tool for search reduction*, in this volume.
8. G. S. Tseitin, *On the complexity of deduction in the propositional calculus*, Zap. Nauchn. Sem. Leningrad. Otdel. Mat. Inst. Steklov. (LOMI) **8** (1968), 234–259; English transl. in Sem. Math. V. A. Steklov Math. Inst. Leningrad **8** (1968).
9. M. Dunham and H. Wang, *Towards feasible solutions of the tautology problem*, Ann. Math. Logic **10** (1976), 117–154.
10. R. Karp, *Reducibility among combinatorial problems*, Complexity of Computer Computations (Proc. Sympos., 1972), Plenum Press, New York, 1972, pp. 85–103.
11. M. Davis and H. Putnam, *A computing procedure for quantification theory*, J. Assoc. Comput. Mach. **7** (1960), 201–215.
12. Z. Galil, *On enumeration procedures for theorem proving and for integer programming*, Automata, Languages, and Programming (Third Internat. Colloq.), Edinburgh Univ., Edinburgh, 1976, pp. 355–381.
13. E. Ya. Dantsin, *Parameters defining the time of tautology recognition by the splitting method*, Semiotics and Information Science, No. 12, VINITI, Moscow, 1979, pp. 8–17. (Russian)
14. _____, *Two tautology proof systems based on the splitting method*, Zap. Nauchn. Sem. Leningrad. Otdel. Mat. Inst. Steklov. (LOMI) **105** (1981), 24–44; English transl. in J. Soviet Math. **22** (1983), no. 3.

15. J. Robinson, *A machine oriented logic based on the resolution principle*, J. Assoc. Comput. Mach. **12** (1965), 23–41.
16. Ch. Chang and R. Lee, *Symbolic logic and mechanical theorem proving*, Academic Press, New York, 1973.
17. P. Yu. Suvorov, *On tautology recognition for propositional formulas*, Zap. Nauchn. Sem. Leningrad. Otdel. Mat. Inst. Steklov. (LOMI) **60** (1976), 197–206; English transl. in J. Soviet Math. **14** (1980), no. 5.
18. G. P. Gavrilov (editor), *Enumeration problems of combinatorial analysis*, Kibernet. Sb., "Mir", Moscow, 1979. (A collection of Russian translations of Western papers. For the contents, see MR 82b:05018.)
19. S. Yu. Maslov, *Information in the calculus and optimization of search.*, Kibernetika (Kiev) **1979**, no. 2, 20–26; English transl. in Cybernetics **15** (1979).
20. S. A. Cook and R. A. Recknow, *The relative efficiency of propositional proof systems*, J. Symbolic Logic **44** (1979), 36–50.
21. M. Evangelist, *Nonstandard propositional logics and their application to complexity theory*, Notre Dame J. Formal Logic **23** (1982), 384–392.
22. Z. Galil, *On the complexity of regular resolution and the Davis-Putnam procedure*, Theoret. Comput. Sci. **4** (1977), 23–46.
23. D. Daugherty, *The complexity of resolution theorem proving*, Abstracts of the ACM, vol. 82C, 1982.
24. R. Statman, *Bounds for proof-search and speed-up in the predicate calculus*, Ann. Math. Logic **15** (1978), 225–287.
25. S. Kleene, *Introduction to Metamathematics*, Van Nostrand, New York, 1952.
26. G. S. Tseitin and A. A. Chubarian, *On some bounds for proof lengths in the classic propositional calculus*, Akad. Nauk Armyan. SSR i Erevan. Gos. Univ.: Mat. Voporosy Kibernet. i Vychisl. Tekhn. **8** (1975), 57–64. (Russian)
27. E. Ya. Dantsin, *On the deductive efficiency of the splitting method*, 3rd All-Union Conf. "Application of Methods of Mathematical Logic", Abstracts of Reports,, Tallinn, 1983, pp. 75–76. (Russian)

Semantics of S. Yu. Maslov's Iterative Method

V. Ya. Kreinovich

ABSTRACT. The article gives several versions of the semantics of the iterative method suggested by S. Yu. Maslov to solve the satisfiability problem. Techniques considered are based on replacing a problem with truth values $\{0, 1\}$ by the problem with the continuous range of truth values; the symmetry requirement (all "continuous" values assigned to the Boolean value 1 are equivalent) uniquely describes a one-parameter class of techniques which contains Maslov's method. The symmetry idea provides a single framework for the modifications of iterative method proposed by S. Yu. Maslov, G. V. Davydov, M. I. Zakharevich, I. V. Melichev, and the author.

1. Statement of the problem

In his papers, Maslov ([1]; for another exposition see [2]), proposed the following iterative algorithm for finding collections satisfying propositional formulas. A formula F is assumed to be reduced to the conjunctive normal form (CNF)

$$\underset{i=1}{\overset{m}{\&}} D_i.$$

In what follows F will always mean such a formula. The idea is to assign to every *literal* x from F (i.e., to every variable a or its negation $\bar{a}$) a real number $\zeta(x)$ called the *deferrence* of the literal (the informal meaning of the deferrence of x is the degree of belief in $\bar{x}$). The mapping ζ that maps every literal x into its deferrence $\zeta(x)$ is called an *obstacle*. Maslov's method starts with assigning some initial values to the deferrence $\zeta^{(0)}$, and then recalculates the obstacles according to the iterative formula

(1.1) $$\zeta^{(n+1)} = T(F)\zeta^{(n)}.$$

In Maslov's original method, $T(F)$ was defined by the formula

(1.2) $$(T(F)\zeta)(x) = L\zeta(x) + R \sum_{\substack{x \in D \\ D \in F}} \min_{\substack{y \in D \\ y \neq x}}(\zeta(y)),$$

where L, R are given positive constants, $D \in F$ means that D is a clause in the initial formula $F = \underset{i=1}{\overset{m}{\&}} D_i$, and $y \in D$ means that the literal y is contained in

©1996 American Mathematical Society

the clause D. A *Boolean vector* is a mapping that associates with each variable in F either 0 (false) or 1 (true). (By arbitrarily numbering the variables contained in F we can turn this definition into the conventional one under which an obstacle is defined as a $2k$-vector consisting of nonzero elements.)

DEFINITION 1.1. An obstacle ζ is said to be *regular* if for any variable a in F one and only one of the relations $\zeta(a) > 0$, $\zeta(\bar{a}) > 0$ holds. We say that a regular obstacle *defines* the Boolean vector b if for all a, $\zeta(a) = 0 \Leftrightarrow b(a) = 1$.

By an *iterative method for recalculating obstacles* we mean an arbitrary mapping T that associates with every cnf F in k variables a continuous mapping $T(F): (R_0^+)^{2k} \to (R_0^+)^{2k}$, where R_0^+ is the set of all nonnegative real numbers.

If we fix an initial obstacle $\zeta^{(0)}$ and such a method, then formula (1.1) defines a sequence of obstacles $\zeta^{(n)}$.

DEFINITION 1.2. An iterative method for recalculating obstacles is said to be *compatible with satisfiability* if the following two properties hold:
1. If ζ is a regular obstacle, b is a Boolean vector defined by this obstacle, and b satisfies the formula F, then the obstacle $(T(F))\zeta$ is also regular and defines the same Boolean vector.
2. If ζ is a regular obstacle and ζ defines a Boolean vector b which is not satisfying for F, then the obstacle $(T(F))\zeta$ does not define b.

Recall that a mapping $T(F): (R_0^+)^{2k} \to (R_0^+)^{2k}$ is called *homogeneous* if $T(\lambda\zeta) = \lambda T(\zeta)$. In what follows we consider only those iterative methods for recalculating obstacles for which the mapping $T(F)$ is homogeneous for all formulas F. In particular, the mapping $T(F)$ corresponding to Maslov's method (1.2) has this property.

Let $|\zeta|$ be an arbitrary norm in R^{2k}. We say that a sequence of vectors $\zeta^{(n)}$ *converges "in the direction"* to the vector ζ if $\lim_n \zeta^{(n)}/|\zeta^{(n)}| = \zeta/|\zeta|$.

THEOREM 1.1. *If an iterative method for recalculating obstacles is compatible with satisfiability, the mapping $T(F)$ is homogeneous for all formulas F, and for some formula F_0 and some initial obstacle $\zeta^{(0)}$ the sequence $\zeta^{(n)}$ (defined by the formula (1.1)) converges "in the direction" to a regular obstacle ζ, then the Boolean vector defined by this obstacle is a satisfying vector for F_0.*

This result is proved along the same lines as the similar result for Maslov's method. Proofs for all theorems can be found in the appendix to this article.

THEOREM 1.2 (Maslov [1, 2]). *Maslov's iterative method (1.1), (1.2) is compatible with satisfiability.*

COROLLARY [1, 2]. *If the obstacles $\zeta^{(n)}$ in Maslov's method converge "in the direction" to a regular obstacle, then the Boolean vector defined by this obstacle is a satisfying one.*

Maslov suggested the following tactics for a practical application of his iterative method: After finitely many iterations, we select a literal x for which the ratio $\zeta^{(n)}/\zeta^{(n)}(\bar{x})$ of deferrences of this literal and of the complementary literal $\bar{x}$ is the smallest (other criteria are also possible). After that, in the formula F, we replace x by "true" and $\bar{x}$ by "false". The resulting formula has one variable less.

Then we apply Maslov's iterations to the simplified formula, etc. The process is repeated until no variables are left in the propositional formula, i.e., until the formula becomes a constant.

If the resulting constant is "false", then we return the answer that the initial formula F is not satisfiable. If this constant is "true", then the formula F is satisfiable. In this case, we can get a satisfying vector for F by assigning to every variable from F the value obtained during the process.

Numerical experiments conducted by Maslov himself have shown that this method does not always lead to a satisfying set, and even if it does, it is often too slow. Therefore, Maslov suggested a modification of his iterative method in which the minimum (see (1.2)) is replaced by the l^p-mean:

$$(1.3) \qquad (T(F)\zeta)(x) = L\zeta(x) + R \sum_{\substack{D \ni \bar{x} \\ D \in F}} \left(\sum_{\substack{y \in D \\ y \neq x}} (\zeta(x))^{-r} \right)^{-1/r}.$$

It can be shown that this method is also compatible with satisfiability.

A natural question arises whether the choice of the iterative method in the form (1.1), (1.2) is accidental. A (negative) answer to this question can be given if we construct a reasonable semantics for Maslov's method. Such a semantics may also help us to find out what modifications of Maslov's method improve its convergence.

In this article, we describe three versions of semantics for Maslov's method: the implicative semantics (Section 2), the extremal semantics (Section 3), and the chemical-kinetic semantics (Section 4).

To be more specific: Section 2.1 describes the concept of implicative semantics (based on the ideas of symmetry), and Section 2.2 includes the precise formulation of the implicative semantics (all the proofs are given in the Appendix). In Section 2.3 the semantics is used to generalize the iterative method. In Section 2.4 the semantics is used to combine the ideas of resolution and extension with Maslov's method, and in Section 2.5 the semantics is used to search for symmetry. In Section 3 the satisfiability problem is formulated as an extremal problem (the same symmetry ideas are used). The gradient method for solving this problem coincides either with Maslov's original method or with the modification of Maslov's method proposed by Melichev. A version of the gradient method leads to another modification of Maslov's method for which we can guarantee that, in some reasonable sense, the quality of the approximation improves after each iteration step. Section 4 includes another—informal—justification of Maslov's method in terms of chemical kinetics. This justification suggests that among all the methods of the type (1.1), the method described by formulas (1.1) and (1.2) is the fastest. This justification also leads to an adaptive modification of Maslov's method for which similar informal arguments suggest that it converges even faster than Maslov's original method.

Appendix 2 explains the results of numerical experiments conducted by Maslov and Kurierov to determine the number of satisfying vectors for random formulas. In Appendix 3, we show that Maslov's operator and its modifications are compatible with satisfiability in a stronger sense than stated in Definition 1.2. The precise formulation of this result is given in Appendix 3; the idea behind this definition is as follows:

In view of Theorems 1.1 and 1.2, a natural tactic for applying Maslov's method is as follows: starting with some initial obstacle $\zeta^{(0)}$, perform iterations until, up to a predefined $\varepsilon > 0$, the process converges to a regular obstacle, i.e., until

$$\left\| \frac{\zeta^{(n)}}{\|\zeta^{(n)}\|} - \frac{\zeta^{(n-1)}}{\|\zeta^{(n-1)}\|} \right\| < \varepsilon,$$

and for each variable a exactly one of the following two inequalities holds:

$$\frac{\zeta^{(n)}(a)}{\|\zeta^{(n)}\|} < \varepsilon, \quad \frac{\zeta^{(n)}(\bar{a})}{\|\zeta^{(n)}\|} < \varepsilon.$$

In this case the Boolean vector b is constructed by setting $b(a) = 1$ if the first inequality is true, and $b(a) = 0$ otherwise.

However, numerical experiments (those conducted by Maslov himself and subsequently by others) have shown that the convergence rate given by this tactic is too slow: the usual situation is that "splitting" along one of the variables a is obtained rather fast, i.e., either $\zeta^{(n)}(\bar{a}) \ll \zeta^{(n)}(a)$, or $\zeta^{(n)}(a) \ll \zeta^{(n)}(\bar{a})$, but many more iterations are often required to obtain such a splitting along all variables. This is why Maslov, for practical applications, suggested a different tactic (described after Theorem 1.2). The natural tactic (to wait for convergence to a regular obstacle) is justified by Theorem 1.2. A question arises whether Maslov's tactic can also be justified by a similar theorem. To be more precise, suppose the iterative process produces an obstacle for which some but not all variables a satisfy the relation $(\zeta(a) = 0 \;\&\; \zeta(\bar{a}) > 0) \lor (\zeta(a) > 0 \;\&\; \zeta(\bar{a}) = 0)$. Can we then set the variable a equal to true if $\zeta(a) = 0$ and false if $\zeta(\bar{a}) = 0$? Or will such a process destroy satisfiability?

Appendix 3 gives an affirmative answer to the question of whether this procedure is acceptable and thereby yields a semantics not only for Maslov's iterative procedure itself, but also for the approach to its practical application suggested by Maslov.

This work was accomplished when the seminar on solving intractable problems was active, and the author expresses his gratitude to all participants of the seminar, especially to G. V. Davydov, I. M. Davydova, E. Ya. Dantsin, M. I. Zakharevich, R. R. Zapatrin, I. V. Melichev, G. E. Mints, S. V. Soloviev, and R. I. Freidzon, for their interest in the work and valuable comments.

2. Implicative semantics

2.1. The idea of semantics. The problem of verifying satisfiability and finding a satisfying set can easily be restated in a numerical form by replacing "false" with 0, "true" with 1, Boolean variables with numerical ones taking values 0 and 1, and the operations $\lor$ and $\&$ with the corresponding operations over numbers. Verification of satisfiability then consists in checking whether the equation

(2.1) $$\underset{i=1}{\overset{m}{\&}} D_i = 1$$

is solvable. If the original propositional formula is satisfiable, then its satisfying vector is a solution to this equation. Such a reduction cannot be directly used to solve the initial problem, since fast methods are known only for equations with continuous variables, while the variables in (2.1) are discrete (namely, their domain is the set $\{0,1\}$). One of the most widely used methods of solving a problem with discrete variables is to first solve a similar problem with continuous variables, and then transform this solution into a solution of the problem with discrete variables.

In order to apply this method to the problem under consideration three steps are required:

(1) Extend the initial domain $\{0,1\}$ to a larger set $A \subset R$, and extend the operations $\vee$ and $\&$ to continuous operations on the set A (there is no need to extend the negation operation since in the problem with continuous variables the literals a and $\bar{a}$ correspond to independent variables). Such an extension should make it possible to write a continuous analogue of the equation (2.1).

Since the logical value of a cnf is affected neither by the order of clauses, nor by the order of literals in every clause, it is natural to require that the continuous analogue of the equation (2.1) does not depend on the order in which the operations $\&$ and $\vee$ are performed, i.e.,

$$(2.2) \qquad p \vee q = q \vee p,$$
$$(2.3) \qquad p \vee (q \vee r) = (p \vee q) \vee r,$$
$$(2.4) \qquad p \ \& \ q = q \ \& \ p,$$
$$(2.5) \qquad p \ \& \ (q \ \& \ r) = (p \ \& \ q) \ \& \ r.$$

(2) Select a method for solving the continuous analogue of (2.1).

(3) Define a mapping $\pi \colon A \to \{0,1\}$ that transforms a solution of the equation (2.1) into a solution of the original equation, i.e., into a satisfying vector.

Let us describe basic ideas for each of the three steps.

Steps (1), (3): extension of the initial domain $\{0,1\}$ to $A \subset R$, extension of $\vee$ and $\&$ to A, and construction of the mapping $\pi \colon A \to \{0,1\}$.

The set A is the result of extending the set $\{0,1\}$. The greater the difference between A and $\{0,1\}$, the more the continuous analogue of equation (2.1) differs from the initial equation, and, consequently, the less hope there is that the solution of the continuous analogue will be close to the solution of the initial equation. It is therefore desirable that the set A differ from $\{0,1\}$ as little as possible (but, of course, making it still possible to go to continuous variables).

An extension A is (in a reasonable sense) minimal if

1. all values corresponding to "false" (i.e., belonging to the set $\pi^{-1}(0)$, which we shall denote by 0_A) are "equivalent" in the sense that there is a transformation taking any of them into any other and preserving the operations $\vee$ and $\&$;
2. the same holds for all values corresponding to "true", i.e., belonging to the set $\pi^{-1}(1)$, which we shall denote by 1_A.

In this case, no property can distinguish between new truth values p and p' that correspond to the same Boolean value, which means that p and p', in effect, represent one and the same value.

In order for the variables to become continuous, the domain must be connected, i.e., it must coincide with an interval (finite or infinite) of the real line.

It is natural to interpret the order relation $p < q$ as "p is less true than q". If the mapping π assigns to an element $p \in A$ the truth value 1 ("true"), and if $q \in A$

is more true than p (i.e., $p < q$), then $\pi(q)$ should also be true. So, the mapping π should be monotonically nondecreasing.

Step (2). The basic idea for an iterative method of solving the continuous analogue of equation (2.1) is as follows: each clause $D = x_1 \vee \ldots \vee x_n$ is equivalent to each of the implications

$$\bar{x}_1 \ \& \ \ldots \ \& \ \bar{x}_{i-1} \ \& \ \bar{x}_{i+1} \ \& \ \ldots \ \& \ \bar{x}_m \to x_i, \tag{2.6}$$

for $i = 1, \ldots, m$. Therefore, if at step n the variables $\bar{x}_1, \ldots, \bar{x}_{i-1}, \bar{x}_{i+1}, \ldots, \bar{x}_m$ are true and so is formula F (and, consequently, so is clause D), it is natural to assume at step $(n+1)$ that x_i is true. Thus, we consider a literal x to be true at step $(n+1)$ if at the preceding step either x was true, or the left-hand side of one of the formulas of the form (2.6) having x as its right-hand side was true. In formal notation:

$$a^{(n+1)} = a^{(n)} \vee \left(\bigvee_{D \ni a} \left(\underset{\substack{x \in D \\ x \neq a}}{\&} \bar{x}^{(n)} \right) \right). \tag{2.7}$$

If we denote by $\zeta^{(n)}(a)$ the degree of belief in $\bar{a}$ at step n, and if we understand $\vee$ and $\&$ in (2.7) as operations on A, we obtain an iterative method for recalculating obstacles which, as we will now show, coincides (under natural conditions and the monotonic behavior described above) with Maslov's method.

2.2. Formulation of the result. 2.2.1 *Extension of the domain of truth values.*

DEFINITION 2.2.1. A *truth structure* is a pair consisting of a connected set A on the real line and a monotonic mapping $\pi \colon A \to \{0, 1\}$.

We will denote $\mathbf{1}_A = \pi^{-1}(1)$ and $\mathbf{0}_A = \pi^{-1}(0)$.

DEFINITION 2.2.2. A mapping $f \colon A \to A$ is said to *preserve the truth structure* if it is monotonically nondecreasing, one-to-one, $f(\mathbf{1}_A) = \mathbf{1}_A$, and $f(\mathbf{0}_A) = \mathbf{0}_A$.

THEOREM 2.1. *Suppose that a truth structure (A, π) satisfies the following conditions*:
  1. *for any $p, q \in \mathbf{1}_A$ there exists a mapping $f \colon A \to A$ which preserves the truth structure and satisfies the relation $f(p) = q$;*
  2. *for any $p, q \in \mathbf{0}_A$ there exists a mapping $f \colon A \to A$ which preserves the truth structure and satisfies the relation $f(p) = q$.*

Then one of the sets $\mathbf{1}_A$, $\mathbf{0}_A$ consists of a single point.

(See Appendix 1 for the proof.)

This result makes it clear why in [1] the analogue of one of the truth values is a single point, while that of the other is an interval. It also explains why in the framework of our semantics it makes no sense to associate with each Boolean variable one variable taking values in A and continue the operation of negation to a continuous function from A to A for which $\bar{\mathbf{1}}_A = \mathbf{0}_A$ and $\bar{\mathbf{0}}_A = \mathbf{1}_A$: in this case each of the sets $\mathbf{0}_A$ and $\mathbf{1}_A$ would consist of a single point (and not just one of them).

In the remaining part of Section 2 we will assume that the single-point set is $\mathbf{0}_A$ (it can be shown that if the single-point set is $\mathbf{1}_A$, the results are similar).

Since we are not using any structure on the real line other than the order $<$, it is natural to consider two truth structures equivalent if one of these structures can be transformed into the other by a monotonic one-to-one mapping of the real line. In particular, we can select this mapping in such a way that the point $\mathbf{0}_A$ is transformed into any given point on the real line. A convenient choice is to make $\mathbf{0}_A$ coincide with 0, i.e., with the initial set of values for "false". Therefore, in what follows, we shall assume that $\mathbf{0}_A = \{0\}$.

According to the meaning of the operations "and" and "or", the degree of belief in the results $A\&B$ and $A\vee B$ of these operations is monotonically nondecreasing in each of the arguments, i.e., as a function of the degrees of belief in the component statements A and B. In addition, "p or 'false'" is the same as p, while "p and 'false'" is always false, i.e., $p \vee 0 = p$ and $p \,\&\, 0 = 0$.

THEOREM 2.2. *Let A be an interval (finite or infinite) one of whose endpoints is 0. Suppose that $\vee, \& : A \times A \to A$ are continuous monotonic operations that satisfy the conditions (2.2)–(2.5), $p \vee 0 = p$, and $p \,\&\, 0 = 0$, and for any $p, q \in A$, $p \neq 0$, $q \neq 0$, there exists a mapping $f\colon A \to A$ preserving the truth structure and the operations $\&$, $\vee$ (in the sense that $f(a \,\&\, b) = f(a) \,\&\, f(b)$, with a similar equality being true for $\vee$) such that $f(p) = q$.*

In this case, modulo a monotonic one-to-one mapping, the set A coincides with $[0, \infty)$ and one of the following four statements holds:
1. $\vee = \max$, $\& = \min$;
2. $\vee = +$, $\& = \min$;
3. $\vee = \max$, $p \,\&\, q = (p^{-1} + q^{-1})^{-1}$;
4. $\vee = +$, $p \,\&\, q = (p^{-r} + q^{-r})^{-1/r}$, where $r > 0$.

Comment: If we substitute such $\vee$, $\&$ into (2.7) and call the truth value of $\bar{a}$ the *deferrence* of the literal a, we obtain formula (1.2) in case (1), formula (1.3) in case (2), and in case (3) the formula dual to (1.2):

$$\zeta^{(n+1)}(x) = \max\left\{\zeta^{(n)}(x), \max_{D \ni \bar{x}} \left(\sum_{y \in D} (\zeta^{(n)}(y))^{-1}\right)^{-1}\right\}.$$

There is no need to consider case (1), since it does not increase the initial set of truth values for literals and therefore gives nothing new as compared with discrete methods for solving the satisfiability problem.

Theorem 2.2 was announced in [3].

REMARK. Case (4) with $r = 1$ is the only one for which there exists a "duality" mapping that transforms $\leq$ into $\geq$ and $\&$ into $\vee$. For this case, Maslov suggested a simple electrical interpretation: the expression for $p \vee q$ coincides with the formula that describes the resistance of a circuit consisting of two serially connected resistances p and q, and the expression for $p \,\&\, q$ coincides with the formula that describes the resistance of a circuit consisting of two parallel resistances p and q.

2.3. Implicative semantics can be used to extend Maslov's method to problems other than satisfiability. In this section we show that the idea of implicative semantics can help to justify not only Maslov's method but also its generalization proposed by Davydov for the following problem:

Suppose we are given k pairwise disjoint finite sets A_i. Each of these sets will be called a *set of alternatives*, and the elements of each set A_i will be called *alternatives*. A *collection of alternatives* b is a mapping that maps each of the sets A_i into its element $b(A_i) \in A_i$. We will consider the problem of finding collections of alternatives that satisfy certain conditions.

EXAMPLE. For a propositional formula F, we can take, as the desired collection of finite sets, the sets $A_i = \{a_i, \bar{a}_i\}$, where the a_i are variables from the formula F. In this case, alternatives correspond to literals, and a collection of alternatives corresponds to a Boolean vector. (Other examples will be given below.)

The definitions of a clause, a formula, a cnf, and a satisfying set can be extended to the case of arbitrary collections of finite sets A_i:

A *clause* is an expression of the form $x \vee y \vee \ldots \vee z$, where $x, y, \ldots, z$ are alternatives. A clause $D \equiv x \vee y \vee \ldots \vee z$ is called *true* for a given set of alternatives b if for some i the alternative $b(A_i)$ coincides with one of alternatives $x, y, \ldots, z$.

A *formula* (*in CNF*) is an expression of the form $F \equiv \&_{p=1}^{d} D_p$, where D_p are clauses. We say that a collection of alternatives b *satisfies* the formula F if for this collection of alternatives, all clauses of the formula F are true.

In this section, we consider methods of solving the following problem: given k sets of alternatives A_i and a formula F in CNF (in the sense of the above definition), does there exist a collection of alternatives satisfying F? If so, how can it be found? This problem will be called the *generalized satisfiability problem*.

Let us describe some specific examples of problems which can be reformulated in this form.

EXAMPLES. (1) *Satisfiability problem for propositional formulas*. As we have already noted, for $A_i = \{a_i, \bar{a}_i\}$ the above definitions of a clause and a formula in CNF coincide with the usual definitions for propositional formulas, and the generalized satisfiability problem becomes the propositional satisfiability problem.

(2) *Job assignment problem* (this and the following example are due to Davydov): k is the total number of people applying for job assignments, A_i are possible assignments for the ith person, d is the number of vacancies, and D_p, $1 \leq p \leq d$, is the condition saying that position p must be filled by one of the applicants qualified for the corresponding job. In this case the existence of job assignments under which every vacancy is filled by a qualified person is equivalent to the satisfiability of the formula $F = \&_{p=1}^{d} D_p$, and the construction of the set of such assignments is equivalent to finding a satisfiable set for F.

(3) *Construction planning problem*: k is the number of sites under construction, A_i is the set of alternative construction plans for site i, d is the number of tasks the sites are supposed to perform after they are constructed, and D_p, $1 \leq p \leq d$, is the list of alternative construction plans for separate sites such that the site is capable of performing task p. In this case the possibility of selecting a collection of construction plans under which all the tasks will be performed is equivalent to the satisfiability of the formula $F = \&_{p=1}^{d} D_p$, while the construction of such a collection of plans is reduced to finding a satisfiable set for F.

Let us describe the iterative method for solving the generalized satisfiability problem. Similarly to the problem of satisfiability of propositional formulas (Section

2.1), every clause $D_p = x_1 \vee \ldots \vee x_m$ is equivalent to each of m implications (2.6); each of these implications, in its turn, is equivalent to the collection of the following implications:

$$(2.8) \qquad \bar{x}_1 \,\&\, \ldots \,\&\, \bar{x}_{i-1} \,\&\, \bar{x}_{i+1} \,\&\, \ldots \,\&\, \bar{x}_m \to \bar{x},$$

for all elements x of the set A_j of all alternatives that contain x_i, except x_i itself. The total number of such implications is $|A_j| - 1$, where $|A_j|$ denotes the number of elements in A_j. Therefore, a natural generalization of formula (2.7) to this case is given by the formula

$$(2.9) \qquad \bar{a}^{(n+1)} = \bar{a}^{(n)} \vee \left(\bigvee_{\substack{b \in A_j \\ b \neq a}} \bigvee_{D \ni b} \left(\underset{\substack{x \in D \\ x \neq b}}{\&}\, \bar{x}^{(n)} \right) \right),$$

where j is such that $a \in A_j$.

Similarly to the case of propositional formulas, the *deferrence* $\zeta(a)$ of an alternative a can be defined as the degree of belief in its negation $\bar{a}$.

DEFINITION 2.1. An *obstacle* is a mapping that maps each alternative a into a nonnegative number $\zeta(a)$. An obstacle is called *regular* if for any i there exists a unique alternative $x \in A_i$ for which $\zeta(x) = 0$. We shall say that a regular obstacle *defines* a collection of alternatives b if $\zeta(b(A_i)) = 0$ for all i.

By replacing $\vee$ and $\&$ in (2.9) with Σ and min, we obtain an algorithm for solving the generalized satisfiability problem. This algorithm is an iterative procedure, in which the values of the obstacles are recomputed by using formula (1.1) with

$$(2.10\text{a}) \qquad (T(F)\zeta)(a) = L\zeta(a) + R \sum_{j: a \in A_j} \sum_{b \in A_j \setminus (a)} M(b),$$

where

$$(2.10\text{b}) \qquad M(b) = \sum_{D: D \ni b} \left(\min_{x: x \in D \setminus (b)} \zeta(x) \right).$$

This expression coincides exactly with Davydov's method (the description of the method, as well as Davydov's results, are presented here with his permission).

If in (2.9) we replace $\bigvee$ by $\sum$, and $a \,\&\, b$ by $(a^{-r} + b^{-r})^{-1/r}$ (i.e., by the l^p-mean), we obtain

$$(2.10'\text{a}) \qquad (T(F)\zeta)(a) = L\zeta(a) + R \sum_{j: a \in A_j} \sum_{b \in A_j \setminus (a)} M'(b),$$

where

$$(2.10'\text{b}) \qquad M'(b) = \sum_{D: D \ni b} \left(\sum_{x: x \in D \setminus (b)} \zeta^{-r}(x) \right).$$

The iterative method (1.1), in which the formula (2.10) is used to recompute the obstacles, will be called the *l^p-modification of Davydov's method*.

For both Davydov's method and its l^p-modification, we can prove results similar to those stated in Section 1 for Maslov's original method.

Definition 1.2 of compatibility of an iterative method with satisfiability admits a natural generalization to the case of arbitrary alternatives.

An iterative method of recalculating obstacles will be called *compatible with satisfiability* if:

1. If ζ is a regular obstacle ζ, and the collection of alternatives b defined by this obstacle satisfies the formula F, then the obstacle $T(F)\zeta$ is also regular and defines the same collection of alternatives.
2. If ζ is a regular obstacle, and the collection of alternatives defined by ζ does not satisfy the formula F, then the obstacle $T(F)\zeta$ does not define b.

In this (more general) case, Theorem 1.1 is also true (with the same formulation and virtually the same proof).

THEOREM 2.3. *Davydov's iterative method* (2.10) *and its l^p-modification* (2.10′) *are compatible with satisfiability.*

This result was proved by Davydov for (2.10).

COROLLARY. *If for Davydov's method or for its l^p-modification the sequence $\zeta(n)$ converges "in the direction" to a regular obstacle ζ, then the collection of alternatives defined by ζ satisfies the formula F.*

In view of this corollary the iterative methods (2.10) and (2.10′) can be used to solve the generalized satisfiability problem and, consequently, to solve problems described in the three examples given above.

Let us describe another important example of applying Davydov's iterative method, viz., the iterative method of Maslov type corresponding to the hypersplitting algorithm (the problem of constructing such a method was formulated by Dantsin).

In order to explain the main idea of the hypersplitting method, let us first recall the main idea of the splitting method. This term describes different deduction tactics that are based on the sequential application of an auxiliary algorithm; this auxiliary algorithm, given a formula F, generates two things: a literal x (from this formula), and a Boolean value ε. If such an auxiliary algorithm is given, then the corresponding "splitting tactic" is defined as the following recursive procedure (recursion is over the number of variables in the formula F):

1. for formulas F that contain no variables (i.e., for constants), the formula is satisfiable if the constant is "true" and not satisfiable if $F =$ "false";
2. if the tactic is already defined for formulas with $\leq k$ variables, then for a formula F with $k+1$ variables the tactic of searching for a deduction is defined as follows:
   (a) apply the given auxiliary algorithm to the formula F; as a result, we get a literal x and a Boolean value ε;
   (b) substitute "true" for x and "false" for $\bar{x}$ in F, and get a formula $F[x]$ with k variables;
   (c) apply the tactic to $F[x]$ (recall that such a tactic is already defined for formulas in k variables);

(d) if the tactic replies that $F[x]$ is satisfiable, output the answer "satisfiable", output as the satisfiable collection the set consisting of x and the satisfiable collection for $F[x]$, and stop the algorithm;

(e) if the tactic replies that $F[x]$ is not satisfiable, and ε is "false", output the answer "F is not satisfiable" and stop the algorithm;

(f) if the tactic declares the formula $F[x]$ to be not satisfiable and ε is "true", apply the described tactic to the formula $F[\bar{x}]$. If this formula $F[\bar{x}]$ is satisfiable, then so is F and the satisfying truth assignment is a collection consisting of $\bar{x}$ and the satisfying collection for $F[\bar{x}]$. If $F[\bar{x}]$ is not satisfiable, then the original formula F is not satisfiable either.

An example of such a tactic is Maslov's tactic described after the corollary to Theorem 1.2: in this case ε is "false", and the literal x is defined as the literal with the largest value of deferrence after a fixed number of iterations. Another example is a so called backtracking, when ε is "true", and the literal x is selected as the first in the sense of some predefined linear ordering.

The hypersplitting method is a natural generalization of the splitting method. In order to formulate the hypersplitting method, we have to define several new notions.

DEFINITION 2.2. A formula $E \equiv E_1 \vee \ldots \vee E_s$ is said to be *admissible* for formula F if the following conditions are satisfied:
1. all variables from E are also contained in F;
2. the formulas E_i are pairwise incompatible (i.e., for every $i \neq j$, E_i and E_j cannot both be true);
3. if F has a satisfying truth assignment, so does F & E.

REMARK. If E is a tautology that satisfies conditions (1) and (2) of Definition 2.2, then E & $F \leftrightarrow F$, and, consequently, E is admissible for F.

DEFINITION 2.3. Let F be an arbitrary formula, and let E be a conjunction of some of the literals from F. By $F[E]$, we will denote a formula obtained from F by substituting "true" for all literals x contained in E and "false" for all the negations of such literals.

The following theorem is an easy consequence of this definition.

THEOREM 2.4 (DANTSIN). *If a formula $E = E_1 \vee \ldots \vee E_s$ is admissible for F, and $F[E_i]$ is not satisfiable for some i, then the formula $E' \equiv E_1 \vee \ldots \vee E_{i-1} \vee E_{i+1} \vee \ldots \vee E_s$ is also admissible for F.*

The main idea of deduction search tactics falling under the general name of the "hyperresolutions method" is as follows. Suppose we are given two algorithms U_1 and U_2. The algorithm U_1 inputs a formula F and generates a formula $E \equiv E_1 \vee \ldots \vee E_s$ that is admissible for F, and a Boolean value ε. The algorithm U_2 inputs F and $E \equiv E_1 \vee \ldots \vee E_s$ and generates $i, 1 \leq i \leq s$. The deduction search algorithm is defined by recursion in the number of variables contained in F:
1. for constants, if F is "true", then F is satisfiable; if F is "false", we return the answer that F is not satisfiable;
2. if the deduction search tactic is already defined for formulas with k or fewer variables, then the tactic for a formula F with $k+1$ variables is defined as follows:

(a) apply the algorithm U_1 to F to obtain $E \equiv E_1 \vee \ldots \vee E_s$ and ε;
(b) apply the algorithm U_2 to F and E to obtain the number i;
(c) apply the deduction search tactics to $F[E_i]$ (recall that for formulas with at most k variables the tactic is already defined);
(d) in case $F[E_i]$ is satisfiable output the answer "F is satisfiable", and for the satisfying truth assignment output the collection consisting of the satisfying collection for $F[E_i]$ and all the literals contained in E_i;
(e) in case $F[E_i]$ is not satisfiable and ε is "false", output the answer "F is not satisfiable" and stop the algorithm;
(f) in case $F[E_i]$ is not satisfiable and ε is "true", then, due to Theorem 2.4, the formula $F' \equiv E_1 \vee \ldots \vee E_{i-1} \vee E_{i+1} \vee \ldots \vee E_s$ is also admissible for F.

If $s = 1$, this means that F is not satisfiable. If $s > 1$, repeat the algorithm described above starting from step (b) but using E' instead of E.

EXAMPLES OF ADMISSIBLE FORMULAS. (1) The formula $a \vee \bar{a}$ is admissible for any formula F. If the algorithm U_1 outputs $E = a \vee \bar{a}$ for all formulas F, the hypersplitting method becomes the splitting method.

(2) As shown in [4], for formulas admitting automorphisms the hypersplitting method can sometimes speed up the search for the satisfying collection as compared to the splitting method. An algorithm U_1 that assigns to every formula F an admissible formula $E = E_1 \vee \ldots \vee E_s$ is described in [4]; however, the problem of constructing the second algorithm U_2 (that would select one of the conjunctive terms E_i) remains open. Let us describe an iterative method of selecting E_i.

The main idea of the method is as follows: introduce s new variables $e_1, \ldots, e_s$, replace F & $(E_1 \vee \ldots \vee E_s)$ by F & $(e_1 \vee \ldots \vee e_s)$ & $(e_1 \leftrightarrow E_1)$ & $\ldots$ & $(e_s \leftrightarrow E_s)$, reduce each of the formulas $e_i \leftrightarrow E_i$ to CNF, and then apply Davydov's method to the resulting formula. Let us give precise definitions.

DEFINITION 2.4. Let $F \equiv \&_{p=1}^{d} D_p$ be a formula in CNF in variables $a_1, \ldots, a_k$. Set $x_i = a_i$ for $i > 0$ and $x_i = \bar{a}_{-i}$ for $i < 0, |i| < k$. Let $E \equiv E_1 \vee \ldots \vee E_S$ (where $E_i = \&_{j=1}^{k_i} x_{n(i,j)}$) be an admissible formula for F. Construct $k+1$ sets of alternative as follows: $A_i = \{a_i, \bar{a}_i\}$ for $i = 1, \ldots, k$, $A_{k+1} = \{e_1, \ldots, e_s\}$. In this case the formula

$$\underset{p=1}{\overset{N_d}{\&}} D_p,$$

where

$$N_d = d + 1 + s + \sum k_i,$$
$$D_{d+1} = e_1 \vee \ldots \vee e_s,$$
$$D_{d+1+i} = e_i \vee x_{-n(i,1)} \vee \ldots \vee x_{-n(i,k_i)},$$
$$D_z = x_{n(i,q)} \vee e_1 \vee \ldots \vee e_{i-1} \vee e_{i+1} \vee \ldots \vee e_s,$$

and $z = d + 1 + s + \sum_{j=1}^{i} k_j + q$, will be called the *alternative selection formula* and denoted by $F * E$.

THEOREM 2.5. *If E is admissible for F, then the formulas F & E and $F * E$ are both satisfiable or not satisfiable simultaneously. In addition, if b is a* satisfying collection for $F * E$, then its restriction to the first k sets of alternatives is a satisfying collection for F.

COROLLARY. *If the result $\zeta^{(n)}$ of applying the iterative method (2.10) or (2.10') to the formula $F * E$ converges "in the direction" to a regular obstacle ζ, and i is such that $\zeta(e_i) = 0$, then the formula $F[E_i]$ is satisfiable simultaneously with F.*

This corollary makes it possible to suggest the following method for selecting the conjunctive term E_i: perform n steps of the iterative method (2.10) or (2.10') for the formula $F * E$, and select i for which $\zeta^{(n)}(e_i)$ is the smallest.

2.4. Resolution and extension rules in Maslov's iterative method. As we have already noted in Section 2.1, the proposed semantics of Maslov's method interprets every clause as a system of implications, and in the framework of the suggested scheme the collection of these implications defines the Maslov's method. An example constructed by Melichev [5] demonstrates that the convergence of Maslov's method can be very slow, but one can speed it up by adding to the system of implications their consequences. In the deduction search theory, the most widely used method of generating such consequences is the resolution method.

If we add to the original implications those obtained from them by applying the resolution rule once, then, replacing $\vee$ and $\&$ with the corresponding functions (as in Section 2.2), we obtain an iterative method which will be called the 2nd order method. The method obtained by adding implications generated by applying the rule of resolutions twice will be called the 3rd order method, and so on.

THEOREM 2.6. *For any l, the lth order method is compatible with satisfiability.*

(The proof is the same as for Theorem 1.2.)

Another way of generating new clauses without changing the satisfiability of the formula is Tseitin's extension rule, which works as follows: for literals x, y, we add a new variable $\alpha(x, y)$, and add to the initial collection of clauses new clauses that form the CNF for the formula $\alpha(x, y) \leftrightarrow x \ \& \ y$. If the formula thus extended is true, then $\alpha(x, y)$ is true or false simultaneously with $x \ \& \ y$.

By applying Tseitin's rule to the pair consisting of $\alpha(x, y)$ and the literal z, we obtain a new variable $\alpha(x, y, z)$ whose value — in case the extended formula is true — coincides with that of $x \ \& \ y \ \& \ z$. After applying the rule p times we obtain variables of the form $\alpha(x, y, \ldots, z)$ that correspond to conjunctions of k literals from the initial formula.

DEFINITION 2.5. Suppose that the collection of clauses D_c is obtained from the initial collection (2.6) by applying the extension rule p times, and then applying the resolution rule l times. The iterative Maslov's method for the formula

$$\& \ D$$
$$D : D \in D_c$$

will be called the (p, l)-*extended Maslov's method*. If D_c is replaced with $D_{\bar{c}}$, where $D_{\bar{c}}$ contains only those clauses from D_c all of whose literals are either of the form $\alpha(x, \ldots, y)$, or of the form $\bar{\alpha}(x, \ldots, y)$ (where $x, \ldots, y$ is a collection of p literals from the original formula), then the corresponding Maslov's method will be called (p, l)-*bounded*.

Since the formula

$$\& \; D$$
$$D \colon D \in D_c$$

is satisfiable or not satisfiable simultaneously with the original formula F, then — in the case where the (p,l)-extended Maslov's method converges "in the direction" to a regular obstacle—the Boolean vector defined by this obstacle provides a satisfying collection for F.

THEOREM 2.7. *Suppose that F is a formula in k variables, $\zeta^{(0)}(\alpha(x,\ldots,y)) = 1$ for all literals $x,\ldots,y$, and $\zeta^{(1)}$ is the result of applying the (k,l)-bounded Maslov's method. Then for all sufficiently large l the following relations hold: $\zeta^{(1)}(\alpha(x,y)) = L$ for satisfying collections $x,\ldots,y$ and $\zeta^{(1)}(\alpha(x,\ldots,y)) \geq R+L$ for all other collections (where R, L are the parameters of Maslov's method).*

COMMENT. One can say that for sufficiently large l the (k,l)-bounded Maslov's method converges in one step. Thus Theorem 2.7 demonstrates that the basic concept underlying methods of Maslov type is complete in the sense that Maslov's method is the first in the family which asymptotically always leads to the solution of the satisfiability problem. Of course, Theorem 2.7 does not make it possible to guarantee an estimate lower than exponential, since for the modification whose completeness we have just proved, each iteration requires at least 2^k computational steps (namely, computing $\zeta(\alpha(x,\ldots,y))$ for all $(x,\ldots,y)$).

2.5. Applying Maslov's method to the search for automorphisms.

According to [4], an automorphism τ of a CNF formula F is a permutation τ of literals in the formula F under which complementary pairs go into complementary pairs, and for every clause $D \equiv x \vee \ldots \vee y$ from the formula F, the clause $\tau(D) = \tau(x) \wedge \ldots \wedge \tau(y)$ is also contained in F. As shown in [4], if a formula admits an automorphism, then it is often easier to find a satisfying collection. This means that it makes sense to find all the automorphisms first.

THEOREM 2.8. *If τ is an automorphism of formula F and $\zeta^{(0)} \equiv 1$, then for all n and x*

(2.11) $$\zeta^{(n)}(x) = \zeta^{(n)}(\tau x).$$

As a corollary, we see that the task of searching for automorphisms is especially important for Maslov's method, because if, for example, $\tau(a) = \bar{a}$ for some variable a, then $\zeta^{(n)}(a) = \zeta^{(n)}(\bar{a})$ for all n, and Maslov's method does not produce a satisfying collection.

Another consequence of the theorem is the justification of the following method of searching for automorphisms. Starting with $\zeta^{(0)} \equiv 1$, perform several steps of Maslov's method. Then consider all permutations τ of the set of literals that preserve complementarity and satisfy relation (2.11), and for each of them check whether this permutation is an automorphism. If there are many such permutations one can use the extension rule. The following theorem holds.

THEOREM 2.9. *Let F be a formula, let τ be its automorphism, let*

$$\zeta^{(0)}(x) = \zeta^{(0)}(\alpha(x,\ldots,y)) = 1,$$

and let $\zeta^{(n)}$ be the result of applying the $(p,0)$-extended Maslov's method. Then

(2.12) $$\zeta^{(n)}(\alpha(x,\ldots,y)) = \zeta^{(n)}(\alpha(\tau x,\ldots,\tau y))$$

for all n and x, y.

Thus, if there are many permutations satisfying (2.11), it is necessary to perform n steps of the $(p,0)$-extended Maslov's method and then consider only those permutations that satisfy (2.12).

3. Extremal semantics

In this section, we show that Maslov's method coincides with the gradient method applied to a certain optimization (extremal) problem that is equivalent to the propositional satisfiability problem.

3.1. In Section 2 we have considered the solution of equation (2.1) by means of a special logical trick. In this Section 3 we will show that a more traditional gradient method leads to the same iterative Maslov method.

If in the original formula $F = \&_{i=1}^{m} D_i$ all literals x are replaced with the corresponding deferrence values $\zeta(x)$, $\wedge$ with min, and $\&$ with $+$, then we obtain the functional

(3.1) $$J(\zeta) = \sum_{D \in F} \min_{x \in D} \zeta(x),$$

where the sum is taken over all clauses D of formula F, and the minimum is taken over all literals in the clause D.

If ζ is a regular obstacle that defines a satisfying collection, then, as one can easily see, $J = 0$. The converse statement is not true: the equation $J = 0$ has a trivial solution $\zeta(a) \equiv 0$. In order to reduce the satisfiability problem to solving an equation we will narrow down the class of obstacles.

DEFINITION 3.1. An obstacle ζ is called *partially regular* if $\zeta(a) + \zeta(\bar{a}) > 0$ for all variables a. We shall say that a partially regular obstacle is *compatible* with the Boolean vector b, if $\zeta(a) = 0 \Rightarrow b(a) = 1$ and $\zeta(\bar{a}) = 0 \Rightarrow b(a) = 0$. (Here if $\zeta(a) > 0$ & $\zeta(\bar{a}) > 0$ for some variable a, then $b(a)$ can be arbitrary.)

THEOREM 3.1. *If ζ is a partially regular obstacle, then $J(\zeta) = 0$ if and only if every Boolean vector that is compatible with ζ satisfies the original propositional formula F.*

Thus the problem of searching for a satisfying vector is reduced to the problem of solving the equation $J = 0$.

The problem of verifying satisfiability can also be restated in terms of J:

THEOREM 3.2. *Satisfiability of a formula F is equivalent to $j = 0$, where j is the minimum of J over all ζ for which $\zeta(a) + \zeta(\bar{a}) = 1$ for any variable a from F. The satisfying collections are exactly those Boolean vectors that are compatible with the obstacle ζ on which J attains its minimum.*

Let us now describe some guiding lines leading to the justification of Maslov's method. The gradient algorithm for minimizing a function L that depends on k

variables $x_1, \ldots, x_k$ consists of the consecutive application of the formula

$$x_i^{(n+1)} = x_i^{(n)} - R \frac{\partial L}{\partial x_i}(x^{(n)_j}).$$

If either L is not differentiable or there are no explicit formulas for its derivatives, one can use, instead of the gradient $\partial L/\partial x_i$, the result of the numerical differentiation

$$x_i^{(n+1)} = x_i^{(n)} - R \frac{DL}{Dx_i},$$

where
(3.2)
$$\frac{DL}{Dx_i} = c(L(x_1^{(n)}, \ldots, x_{i-1}^{(n)}, x_{i+1}^+, \ldots, x_k^{(n)}) - L(x_1^{(n)}, \ldots, x_{i-1}^{(n)}, x_{i+1}^-, \ldots, x_k^{(n)}))$$

for some c, x_i^-, and x_i^+. In the case under consideration the functional J is minimized over the set of all ζ for which $\zeta(x) + \zeta(\bar{x}) = 1$ for any literal x. Therefore, $\partial J/\partial \zeta(x) = -\partial J/\partial \zeta(\bar{x})$. Since the function J is not everywhere differentiable, the gradient method takes the form

(3.3) $$\zeta^{(n+1)}(x) = \zeta^{(n)}(x) + R \frac{DJ}{D\zeta(\bar{x})}.$$

In order to minimally deviate from the original discrete problem, it is natural to use in the definition of numerical differentiation the values $\zeta^-(x_i) = 0$ (which corresponds to "true") and $\zeta^+(x_i) = 0$ (which corresponds to "false").

THEOREM 3.3 (Zakharevich). *For the function defined by relations* (3.2), (3.3) *with* $\zeta^-(x_i) = 0$ *and* $\zeta^+(x_i) = \infty$, *the gradient method coincides with Maslov's method.*

If $\zeta^+(x_i) = \zeta^{(n)}(x_i)$, then formula (3.3) leads to the following modification of Maslov's method:

(3.4) $$\zeta^{(n+1)}(x) = \zeta^{(n)}(x) + R \sum_{D \ni \bar{x}} \min_{y \in D}(\zeta^{(n)}(y)).$$

This modification was proposed by Melichev [5] to speed up the convergence of Maslov's method for 2-CNF formulas.

3.2. Extremal semantics helps to generalize Maslov's method to problems other than satisfiability. In Section 2 it was shown that the implicative semantics can be used to extend Maslov's method to the generalized satisfiability problem. In the present section we will show that the same generalization of Maslov's method, namely, Davydov's method, can also be obtained by using the extremal semantics.

The problem solved by Davydov's method was stated in Section 2; in this section, we will use the notation and definitions from Section 2.

DEFINITION 3.2. We say that an obstacle ζ is *partially regular* if for each i there exists at most one alternative $a \in A$ for which $\zeta(a) = 0$. A partially regular obstacle ζ is called *compatible* with the collection of alternatives b if the conditions $\zeta a = 0$ and $a \in A_i$ imply that $a = b(A_i)$.

THEOREM 3.5. *If ζ is a partially regular obstacle, then $J(\zeta) = 0$ if and only if every collection of alternatives b that is compatible with ζ satisfies the original formula F.*

THEOREM 3.6. *A formula $F \equiv \& \; D_i$ is satisfiable if and only if $j = 0$, where j is the minimum of the functional J (defined by the formula (3.1)) over all obstacles ζ for which for all i, $\sum_{a \in A_i} \zeta a = 1$. Here satisfying collections are exactly those collections that are compatible with the obstacle ζ on which J attains its minimum.*

Since $\sum_{a \in A_i} \zeta(a) = 1$, we have

$$-\frac{\partial J}{\partial \zeta(a)} = \sum_{\substack{b \neq a \\ b \in A_j}} \frac{\partial J}{\partial \zeta(b)}$$

for all $a \in A_j$, and, consequently, the gradient method can be defined by the following formula:

$$(3.5) \qquad \zeta^{(n+1)}(a) = \zeta^{(n)}(a) + R \sum_{\substack{b \neq a \\ b \in A_j}} \frac{DJ}{D\zeta(b)}.$$

THEOREM 3.7. *The gradient method defined for the functional J by formulas (3.5), (3.2) with $\zeta^+ = \infty$, $\zeta^- = 0$ coincides with Davydov's method.*

3.3. The extremal semantics can be used to *speed up the convergence* of the original Maslov's method by selecting R at each step in such a way that the minimizing functional J gets smaller after each iteration of Maslov's method. To achieve this, we *normalize* ζ, i.e., go to $\zeta'(x) = \zeta(x)/(\zeta(x) + \zeta(\bar{x}))$, and substitute the normalized ζ' into J. If the value of J obtained after the $(n+1)$th iteration is greater than the value after the nth iteration, change R until J becomes smaller (the same idea can be applied to the methods of Melichev, Davydov, and others).

4. Justification of Maslov's method in terms of chemical kinetics

The idea of a relationship between some models of chemical kinetics and the satisfiability problem belongs to Matiyasevich [5]. We will show that this idea leads to another justification of Maslov's method.

Let F be a formula with k variables $a, \ldots, b$. With each literal $a, \bar{a}, \ldots$ we associate a substance denoted by the same letter. The concentration of a substance x is proportional to the degree of belief in the corresponding literal x. Each implication in (2.6) is interpreted in the following sense: the presence of substances $\bar{x}_1, \ldots \bar{x}_{i-1}, \bar{x}_{i+1}, \ldots, \bar{x}_m$ leads to the appearance of substance x_i. Here the degree of belief in the literals is nondecreasing, therefore so is the concentration of substances $\bar{x}_j$. Thus, the increase in the concentration of substance $\bar{x}_i$ cannot be explained by a decrease in the concentrations of other substances x_j and $\bar{x}_j$. Thus, to make this chemical "model" realistic, we must have a background substance X that goes into $\bar{x}_i$. In chemical terms, each substance x_j serves as a catalyst for the reaction $X \to x_i$. To be more precise, we have the reaction

$$(4.1) \qquad \bar{x}_1 + \ldots + \bar{x}_{i-1} + \bar{x}_{i+1} + \ldots + \bar{x}_m + X \to \bar{x}_1 + \ldots + \bar{x}_m + x_i.$$

Let us describe the correspondence between the satisfiability problem and this chemical kinetics model.

DEFINITION 4.1. Let $S = \{a_1, \ldots, a_n\}$ be a finite set; its elements will be called *substances*. A *state* of the system of substances is a nonnegative n-vector n_i; the ith coordinate of the state vector n_i is called the *concentration* of the substance a_i. A *reaction* between substances in S is a triple consisting of two integer nonnegative n-vectors $\vec{k}(s) = (k_1, \ldots, k_n)$, $\vec{l}(s) = (l_1, \ldots, l_n)$, and of a positive number $c(s)$. The reaction is usually written in the following way:

$$(4.2) \qquad \sum k_i(s) a_i \stackrel{c(s)}{\to} \sum l_i(s) a_i.$$

(Terms with $k_i = 0$ and $l_i = 0$ can be omitted.) The sum $\sum k_i a_i$ is called the *input* of the reaction, $\sum l_i a_i$ its *output*, and $c(s)$ the *intensity* of the reaction.

The speed of the reaction $v(s)$ depends on its intensity and on the concentration of the substances taking part in the reaction. If the concentrations are small, then, for a reaction to take place, the molecules of all the substances have to meet. The probability of such an encounter is proportional to all the concentrations, and therefore, in this case, the reaction rate v is proportional to the product of the concentrations:

$$(4.3) \qquad v(s) = c(s) \prod_{i=1}^{n} n_i^{k_i(s)}.$$

The greater the concentrations, the faster the reactions. In the case of high concentrations (and, therefore, fast reactions) the reaction always takes place provided all the necessary substances are present. The reaction requires $k_i(s)$ molecules of each substance a_i. The total number of such collections of k_i molecules is proportional to $n_i / k_i(s)$; therefore, in this case, the reaction rate is proportional to the *minimum* of the numbers $n_i / k_i(s)$:

$$(4.4) \qquad v(s) = c(s) \cdot \min_{i : k_i(s) > 0} \left\{ \frac{n_i}{k_i(s)} \right\}.$$

DEFINITION 4.2. The *speed of a slow reaction* is the number defined by (4.3). The *speed of a fast reaction* is the number defined by (4.4).

The reaction s changes the concentrations according to the formula

$$(4.5) \qquad \dot{n}_i = -(k_i(s) + l_i(s)) v(s).$$

DEFINITION 4.3. A *system of reactions* is a pair consisting of a finite set S and a finite collection R of reactions between substances from S.

The *dynamic system* corresponding to a system of reactions is the system of equations

$$(4.6) \qquad \dot{n}_i = \sum_{s \in R} (l_i(s) - k_i(s)) v(s).$$

A *difference scheme* for solving a dynamic system (4.6) is the iterative process:

$$n_i^{(n+1)} = n_i^{(n)} + \dot{n}_i(n_i^{(n)}) \cdot \Delta t.$$

THEOREM 4.1. *A difference scheme for a dynamic system corresponding to the system of fast reactions* (4.1) *with equal intensities $c(s)$ coincides with Maslov's iterative method* (1.1).

Thus, chemical kinetics leads to a new semantics for Maslov's method under which R is interpreted as $c\Delta t$, where Δt is the discretization step of the difference scheme.

REMARK. An interpretation of Maslov's scheme as a difference scheme for a system of differential equations (but without providing a specific form of this scheme) was proposed independently by Melichev.

What are the new implications of this semantics?
(1) *An heuristic argument in favor of selecting* min *in Maslov's method in order to achieve the fastest convergence.* This is related to the fact that min corresponds to the fastest chemical reactions.

There is another argument (a somewhat weaker one) in favor of the statement that an iterative method of Maslov's type is the best method of searching for a satisfying vector. Namely, information processing in the human brain is performed by chemical reactions inside neurons. Since a human being is a product of long progressive evolution, our way of processing information (which is described by equations of chemical kinetics) must be optimal or close to optimal. Maslov's method is a good approximation for chemical kinetics of fast reactions, and so Maslov's method must also be close to optimal.

(2) *An heuristic argument for selecting the best parameter in Maslov's method.* R is proportional to the step Δt of the difference scheme. Let T be the time after which the reactions stop. The smaller Δt, the more steps ($\approx T/\Delta t$) are required by the algorithm. However, one cannot take Δt too large: otherwise, $n_i(t) + \dot{n}_i\Delta t$ will be a bad approximation for $n_i(t+\Delta t)$. Therefore, it makes sense to take as Δt the largest value of Δt for which the error of this approximation is not large. This error is of the form

$$n_i(t+\Delta t) - n_i(t) - \dot{n}_i(t) \approx \ddot{n}_i(t)\Delta t^2,$$

hence the condition that this error is relatively small takes the form

$$\left\|\frac{1}{2}\ddot{\vec{n}}(t)\Delta t^2\right\| \leq \|\dot{\vec{n}}(t)\Delta t\|,$$

where $\vec{n}$ is the state vector. Therefore, the largest Δt for which this condition is satisfied is equal to

(4.7) $$R \approx \Delta t \approx 2\frac{\|\dot{\vec{n}}(t)\|}{\|\ddot{\vec{n}}(t)\|}.$$

For our reaction system, differentiating the formula for $\dot{n}_i(t)$ with respect to t, we obtain the following expression for $\ddot{n}_i(t)$:

(4.8) $$\ddot{x}(t) = \sum_{j:\bar{x}\in D_j} \dot{x}^j,$$

where the literal x^j is defined by the condition

$$x^j = \min_{x: x \in D_j} x.$$

This formula can be used in two different ways: we can either recalculate R by using formula (4.7) at each step of Maslov's method, or else, for each class of propositional formulas, we can compute R (by formula (4.7)) for several typical formulas from this class, take the average of these values of R, and use this average value for all formulas from the given class.

(3) *Potential modifications of Maslov's method.* In terms of the semantics of chemical kinetics, Maslov's method is generated by the system of reactions (4.1), which are assumed to have equal intensities. If we are not happy with the original Maslov's method, we can try to modify it by changing the intensities of the reactions (4.1), and/or by adding new reactions.

(3.1) *"Setting to zero"*. In Section 1 a tactic for practical application of Maslov's iterative method is described; this tactic includes "setting to zero" the obstacles $\zeta(x)$, where x is a literal such that $\zeta(x) = \max \zeta(x)$. Such "setting to zero" can be included in the iterative process if one adds to the chemical reactions (4.1) the "annihilation" of substances a and $\bar{a}$, i.e., the reaction

(4.9) $$a + \bar{a}^N \to X + X.$$

This is equivalent to adding the term $N \min(\zeta(x), \zeta(\bar{x}))$ to the right-hand side of (1.2). The corresponding method will be called the *A-modification* of Maslov's method. Since for regular obstacles this new term is equal to 0, and since the compatibility condition uses the value for $T\zeta$ only for regular obstacles ζ, Theorem 1.2 can be easily extended to A-modifications; in other words, the following theorem holds.

THEOREM 4.2. *A-modification of Maslov's method is compatible with satisfiability.*

Therefore, if iterations converge "in the direction" to a regular obstacle, then this obstacle defines a satisfying collection.

(3.2) *A modification whose goal is to speed up the convergence.* In order to achieve such a modification, one can adjust the speed of different reactions, i.e., make R in (1.1) dependent on D and n, and N dependent on a and n. This dependency can, for example, be defined as follows: if the sum $\zeta(a) + \zeta(\bar{a})$ is increasing, then one has to increase $N(a, n)$; if this sum is decreasing, one has to decrease $N(a, n)$. If the difference $|\zeta(a) - \zeta(\bar{a})|$ does not grow fast enough, one has to increase $R(D, n)$ for those D which contain either a or $\bar{a}$.

DEFINITION 4.4. The *adaptive Maslov's method* is the method defined by the formula
(4.10)
$$\zeta^{(n+1)}(x) = L\zeta^{(n)}(x) + \sum_{D: D \ni \bar{x}} R(D,n)(\min_{y: y \in D \setminus (\bar{x})} \zeta^{(n)}(y)) - N(x,n) \min(\zeta^{(n)}, \zeta^{(n)}(\bar{x}))$$

and the rules

$$R(D, n+1) = f(R(D,n), \zeta^{(n)}),$$
$$N(x, n+1) = g(N(x,n), \zeta^{(n)}).$$

THEOREM 4.3. *The adaptive Maslov's method is compatible with satisfiability.*

(3.3) *Modification of Maslov's method for solving a similar informal problem.* Suppose that we have to find a solution to some creative informal problem, for which the desired solution can be naturally described in formal terms, i.e., as a word in an appropriate alphabet. However, the condition that this word must satisfy is not formal: e.g. a verse must be beautiful, or a musical sentence must express sorrow. Various formalizations of this informal condition can, of course, be described, but each of them will be adequate only to a certain degree; this degree can, for example, be estimated numerically as the relative number of experts who consider this formalization to be adequate. Since the satisfiability problem is a universal search problem, all formalizations of the original problem can be reduced to the satisfiability problem. In other words, instead of searching for a desired word, one can search for a Boolean vector b, and express each of the formalizations by a formula F_i in CNF which has to be satisfied by b. To each of these formalizations, we assign the degree of belief $\chi(F_i)$ that this formalization is adequate. The semantics of chemical kinetics suggests that in order to solve the original informal problem, we consider the system of all reactions of the form (4.1) that correspond to all formulas F_i, with reaction rates proportional to $\chi(F_i)$.

The corresponding operator is of the form

$$(T\zeta)(x) = L\zeta(x) + R \sum_i \chi(F_i) \cdot T_i(x),$$

where

$$T_i(x) = \sum_{D: D \in F_i \ \& \ D \ni \bar{x}} (\min_{y: y \in D \setminus (\bar{x})} \zeta(y)).$$

Appendix 1. Proofs

THEOREM 1.1. By definition, the sequence of obstacles satisfies relation (1.1). The homogeneity condition implies that

(A1) $$T(F_0)\left(\frac{\zeta^{(n)}}{\|\zeta^{(n)}\|}\right) = \frac{(T(F_0)\zeta^{(n)})}{\|\zeta^{(n)}\|},$$

and due to condition (1.1)

(A2) $$T(F_0)\left(\frac{\zeta^{(n)}}{\|\zeta^{(n)}\|}\right) = \frac{\zeta^{(n+1)}}{\|\zeta^{(n)}\|} = \left(\frac{\zeta^{(n+1)}}{\|\zeta^{(n+1)}\|}\right) \cdot \left(\frac{\|\zeta^{(n+1)}\|}{\|\zeta^{(n)}\|}\right).$$

By normalizing the left- and right-hand sides of (A2) we obtain

(A3) $$\frac{T(F_0)\left(\frac{\zeta^{(n)}}{\|\zeta^{(n)}\|}\right)}{\left\|T(F_0)\left(\frac{\zeta^{(n)}}{\|\zeta^{(n)}\|}\right)\right\|} = \frac{\zeta^{(n+1)}}{\|\zeta^{(n+1)}\|}.$$

By the condition of the theorem, $\zeta^{(n)}/\|\zeta^{(n)}\| \to \zeta/\|\zeta\|$. Therefore, due to the continuity of $T(F_0)$, we have $T(F_0)(\zeta^{(n)}/\|\zeta^{(n)}\|) \to T(F_0)(\zeta/\|\zeta\|)$; hence, by going to the limit in (A3), we obtain

$$\text{(A4)} \qquad \frac{T(F_0)\left(\frac{\zeta}{\|\zeta\|}\right)}{\left\|(T(F_0)\frac{\zeta}{\|\zeta\|})\right\|} = \frac{\zeta}{\|\zeta\|}.$$

Now, the homogeneity of $T(F)$ implies that

$$\text{(A5)} \qquad T(F_0)\zeta = \zeta \cdot \frac{\|T(F_0)\zeta\|}{\|\zeta\|}.$$

The right-hand side of (A5) defines the same Boolean vector as ζ; therefore, the definition of compatibility with satisfiability implies that the regular obstacle ζ defines a satisfying collection.

THEOREM 2.1. Since π is monotonic, both $\mathbf{0}_A$ and $\mathbf{1}_A$ are connected sets (i.e., either single points, or intervals), and every point from $\mathbf{0}_A$ is smaller that any point from $\mathbf{1}_A$. The boundary p between $\mathbf{0}_A$ and $\mathbf{1}_A$ lies between the points of $\mathbf{0}_A$ and $\mathbf{1}_A$, i.e., between points of A. Since A is connected, it lies in A. The fact that $A = \mathbf{0}_A \cup \mathbf{1}_A$ implies that either $p \in \mathbf{0}_A$ or $p \in \mathbf{1}_A$. Every monotonic one-to-one function $f: R \to R$ is continuous; hence the relations $f(\mathbf{0}_A) = \mathbf{0}_A$ and $f(\mathbf{1}_A) = \mathbf{1}_A$ imply that $f(p) = p$. This means that such a function f cannot map p into any other point, so the set ($\mathbf{0}_A$ or $\mathbf{1}_A$) which contains p must consist of a single point.

THEOREM 2.2. (1) By an argument similar to that used in the proof of Theorem 2.1, we can show that the set A has no second boundary point; hence there exists a monotonic transformation mapping A onto $[0, \infty)$. We will assume that $A = [0, \infty)$.

(2) In algebraic terms (2.2) and (2.3) mean that $(A, \vee)$ is a commutative semigroup. The fact that $\vee$ is continuous means that it is a topological semigroup, while the fact that $\vee$ is monotonic means that $(A, \vee, \leq)$ is an ordered semigroup. The mapping f defined in the statement of the theorem is an automorphism of an ordered semigroup. Therefore, the theorem is equivalent to the statement that the semigroup is *homogeneous* in the sense that the group of automorphisms of the ordered semigroup $[0, \infty)$ is transitive on $[0, \infty)$.

The following statement holds:

LEMMA 1. *If $(M, *, \leq)$ is a homogeneous commutative ordered topological group with 0, and M is isomorphic to R_0^+ both as an ordered set with 0 and as a topological space, then M is isomorphic to either $(R_0^+, +)$ or $(R_0^+, \max)$.*

(The proof is given below.)

We will now show that the statement of the theorem follows from the lemma. Indeed, applying the lemma to the case where $M = A$, $* = \vee$, we see that modulo a monotonic one-to-one correspondence, either $\vee = +$ or $\vee = \max$. In the first case the automorphisms are given by the transformations of the form $f(x) = kx$, in the second by all monotonic functions f.

In a similar manner one can show that the operation & coincides (modulo a decreasing one-to-one correspondence g) either with $+$ or with $\max$, i.e., either

$g(p \ \& \ q) = gp + gq$, in which case $p \ \& \ q = g^{-1}(gp + gq)$, or $g(p \ \& \ q) = \max(gp, gq)$, in which case, since g is monotonically decreasing, we have $p \ \& \ q = \min(p, q)$.

If $\vee = \max$, $\& = \min$, we obtain case 1 of the theorem, and we obtain case 2 if $\vee = +$, $\& = \min$. For $\vee = \max$ and $p \ \& \ q = g^{-1}(gp+gq)$, a monotonic one-to-one mapping $p \to \frac{1}{g(p)}$ can reduce the situation to case 3. The only remaining case is $\vee = +$ and $p \ \& \ q = g^{-1}(gp + gq)$. In this case, by the condition of the theorem, for every $p, q \in A$ there exists a mapping f preserving $\vee$ and $\&$. Since f preserves $\vee$, we conclude that $fx = kx$, and, therefore, the fact that $\&$ is invariant under f means that for all k, p, q we have

$$g^{-1}(g(kp) + g(kq)) = kg^{-1}(gp + gq),$$

i.e., if $gp + gq = gr$, then $g(kp) + g(kq) = g(kr)$. Therefore the mapping $gp \to g(kp)$ is both additive and continuous, hence linear, i.e., $g(kp) = c(k)g(p)$ for some function $c(k)$. Thus $g(kp) = c(k)g(p) = c(p)g(k)$ and $g(p)/c(p) = g(k)/c(k)$ for all k, p, i.e., $g(p)/c(p) = \text{const}$. Hence $g(kp) = \text{const} \, g(k)g(p)$, and so $g(p) = Ap^r$ for some A and r, i.e., we have case (4). Therefore, to prove the theorem it is sufficient to prove the lemma.

PROOF OF THE LEMMA. (1) If $p * p = p$ for some $p \neq 0$, then due to homogeneity this relation holds for all p. For any $p \leq q$ we have $0 \leq p \leq q$; hence the fact that $*$ is monotonic implies that $0 * q \leq p * q \leq q * q = q$. Since 0 is the zero of the semigroup, we have $0 * q = q$; hence $p * q = q = \max(p, q)$.

(2) Now consider the case where $p * p \neq p$ for all $p \neq 0$. An isomorphism F that maps $(R_0^+, +)$ onto M will be constructed in several steps. At each step we shall check that the mapping is monotonic and one-to-one.

(2.1) Set $F(1) = 1$ and $F(n) = 1 * \cdots * 1$ (n times). Since M is a semigroup, we have $F(n + m) = Fn * Fm$. The conclusion that F is monotonic follows from the same property of $*$ and from the properties of 0:

$$F(n+1) \stackrel{\text{def}}{=} 1 * F(n) \geq 0 * F(n) = F(n).$$

One-to-one property: suppose that $F(n) = F(n+1)$. Then

$$F(n+2) = 1 * F(n+1) = 1 * F(n) = F(n+1) = F(n),$$

$F(2n) = F(n) * F(n) = F(n)$, while we are considering the case where $p * p \neq p$ for all p. Therefore $F(n) \neq F(n+1)$, and the monotonic condition implies that $F(n) < F(n+1) < F(n+2) < \ldots < F(n+k)$, i.e., $F(n) \neq F(n+k)$ for all $k > 0$.

(2.2) Since M is homogeneous, there exists an automorphism f mapping $F(2)$ into $F(1)$. Set $F(2^{-(n+1)}) = f(F(2^{-n}))$. Then the relation $F(1) * F(1) = F(2)$ and the fact that f is an automorphism imply that $F(p \cdot 2^{-q}) = F(2^{-q}) * \ldots * F(2^{-q})$ (p times). The mapping F is monotonic and one-to-one.

(2.3) For any $\alpha \in R_0^+$ set

$$F(\alpha) = \inf\{F(p \cdot 2^{-q}) \mid p \cdot 2^{-q} \geq \alpha\}.$$

Let us prove that F is monotonic. Indeed, if $\alpha < \beta$, then there are rational points between α and β whose denominators are powers of 2: $\alpha < p \cdot 2^{-q} < (p+1) \cdot 2^{-q} < \beta$; therefore $F(\alpha) \leq F(p \cdot 2^{-q}) < F((p+1) \cdot 2^{-q}) \leq F(\beta)$, whence

$F(\alpha) < F(\beta)$. Thus F is a monotonic one-to-one mapping of R_0^+ onto $F(R_0^+) \subset M$. In order to prove that F is a desired isomorphism, it is sufficient to show that $F(R_0^+) = M$, i.e., that for every $p \in M$ there exists β such that $F(\beta) = p$.

(2.4) Let us prove that this relation is satisfied for

$$\beta = \sup\{r \mid F(r) < p\}.$$

(2.4.1) Let us show that $\beta > 0$. It is sufficient to prove that $F(2^{-n}) \to 0$ as $n \to \infty$. (Indeed, then $F(2^{-n}) < p$ for sufficiently large n, and, therefore, $\beta \geq 3^{-n} > 0$.) This can be shown easily: $F(2^{-n})$ decreases monotonically, and so there exists $l = \lim F(2^{-n})$. Since $F(2^{-(n+1)}) * F(2^{-(n+1)}) = F(2^{-n})$, we have in the limit $l * l = l$, and since we consider the case where $p * p = p$ for $p \neq 0$, we have $l = 0$.

(2.4.2) A similar argument proves that $F(2^n) \to \infty$ and $\beta < \infty$.

(2.4.3) In view of the definition of β and the fact that F is monotonic, we have $F(r) < p$ for $r < \beta$ and $F(r) \geq p$ for $r > \beta$. In particular, $x_n < p \leq y_n$, where $x_n = F(\beta - 2^{-n})$, $y_n = F(\beta + 2^{-n})$.

The sequences x_n and y_n are monotonic and, therefore, have the limits x and y. By going to the limit in the inequality $x_n < p \leq y_n$, we conclude that $x \leq p \leq y$. Since $y_n = x_n * F(2 \cdot 2^{-n})$, we have in the limit case $y = x * 0 = x$, whence $y = x = p$. On the other hand, the fact that F is monotonic implies that $x_n < F(\beta) < y_n$, and so $y = x = F(\beta)$ and $F(\beta) = p$. The statement is proved, and so is the lemma. The proof of Theorem 2.2 is complete.

THEOREM 2.3. Suppose that a regular obstacle ζ defines a satisfying collection b. Let us prove that $T\zeta$ is a regular obstacle defining the same satisfying collection. Indeed, if for some alternative a the relation $\zeta a = 0$ is satisfied, then a is contained in a satisfying collection b. Therefore, if $a \in A_i$ and $c \in A_i, c \neq a$, then c is not contained in this collection. Since b is a satisfying collection, the formula F becomes true after the substitution of b, and so all its clauses are true, i.e., each clause contains alternatives from this collection. If an arbitrary clause D contains c, then, since c is not contained in b, there exists $x \in D$, $x \neq c$, such that x belongs to a satisfying collection, i.e., $\zeta x = 0$. Thus

$$\min_{x \in D, x \neq c} \zeta x = 0$$

for all D and c such that $D \ni c$, $c \neq a$, $a \in A_i$.

Substituting this in (2.10), we conclude that

$$(T(F)\zeta)(a) = L\zeta(a) + R \cdot 0 = 0.$$

If $\zeta a > 0$, then $(T(F)\zeta)(a) \geq L\zeta a > 0$. Thus $\zeta a = 0 \Leftrightarrow (T(F)\zeta)(a) = 0$, i.e., $T(F)\zeta$ is a regular obstacle defining the same satisfying collection as ζ.

Suppose now that the collection of alternatives b defined by the regular obstacle ζ is not a satisfying one. This means that there exists a clause D none of whose alternatives belongs to this collection. Let c be any alternative in D, $c \in A_i$, $c \neq b(A_j)$. Then

$$\min_{x \in D, x \neq c} \zeta x > 0;$$

hence $(T(F)\zeta)(b(A_j)) > 0$. If $c \in A_i, c \neq b(A_i)$, then $(T(F)\zeta)(c) > 0$, i.e., $T(F)\zeta$ is not a regular obstacle.

The proof for (2.10') goes on similar lines.

THEOREM 2.4. Since by the conditions of the theorem E is a tautology, any satisfying collection for F is also a satisfying collection for $F \ \& \ E$.

THEOREM 2.5. Conditions (1) and (2) are satisfied for F and E'. If b is a satisfying collection for F, then b is also a satisfying collection for $F \ \& \ E \Leftrightarrow (F \ \& \ E_i) \vee (F \ \& \ E')$. By the hypothesis the formula $F \ \& \ E_i$ is not satisfiable; therefore this collection satisfies $F \ \& \ E'$, i.e., condition (3) of Definition 2.2 holds.

THEOREM 2.6. (1) The condition $e_i \leftrightarrow E_i$ is equivalent to the formula

$$E_i^* \equiv (x_{n(i,1)} \vee \bar{e}_i) \ \& \ \ldots \ \& \ (x_{n(i,k_i)} \vee \bar{e}_i) \ \& \ (e_i \vee \overline{x_{(n(i,1)}} \vee \ldots \vee \overline{x_{n(i,k_i)}}).$$

Since the E_i are pairwise noncompatible, then, if E is true, each of the formulas $\bar{e}_i$ is equivalent to $e_1 \vee \ldots \vee e_{i-1} \vee e_{i+1} \vee \ldots \vee e_s$. Substituting this expression in E_i^* and combining all the clauses of formula F and the formulas obtained, we arrive at the formula $F * E$.

(2) Let b be a satisfying collection for $F \ \& \ E$. Then due to (1), for any e_i equal to the truth value of E_i this collection b satisfies all clauses of the formula $F * E$, and therefore satisfies the formula $F * E$ itself.

(3) Let b be a satisfying collection for $F * E$ and let i be such that $b(A_{k+1}) = e_i$. In this case all the clauses of D_P are satisfied, including those for which $p \leq d$. Hence so is F, and, due to (1), so is the formula $e_i \leftrightarrow E_i$; therefore, so is E_i. Hence the restriction of the collection b to A_j, $j \leq k$, is a satisfying collection for $F[E_i]$.

THEOREM 2.7. The collection of k literals $x, \ldots, y$ describes a unique Boolean vector b. For the sake of brevity we shall denote $\alpha(x, \ldots, y)$ by $\alpha(b)$. Since the resolution method is complete, for l sufficiently large the list of implications obtained in k steps coincides with the list of all possible implications of the form $\bar{\alpha}(b_1) \ \& \ \ldots \ \& \ \bar{\alpha}(b_s) \rightarrow \bar{\alpha}(b)$ following from F. If b is a satisfying collection, then this implication (equivalent to $\alpha(b) \rightarrow \alpha(b_1) \vee \ldots \vee \alpha(b_s)$), which, due to the relation $\alpha(x, \ldots, y) \leftrightarrow x \ \& \ \ldots \ \& \ y$ implies noncompatibility of $\alpha(b)$ for different b, holds only in the trivial case $b = b_1 \vee \ldots \vee b = b_s$. Hence $\zeta^{(1)}(\alpha(b)) = L$.

If b is an unsatisfying collection, then $\overline{\alpha(b)}$ is true, and therefore in the list of implication there are implications having $\overline{\alpha(b)}$ in the right-hand side. The contribution of each of them to the right-hand side of (1.2) is $\geq R$; hence $\zeta^{(1)}(\alpha(b)) \geq R+L$. The theorem is proved.

THEOREM 2.8. The sequence $\zeta^{(n)}$ is defined by the recursive scheme $\zeta^{(n+1)}(x) = (T\zeta^{(n)})(x)$. Therefore, in order to prove (2.11), it is sufficient to show that if some obstacle ζ satisfies $\zeta x = \zeta(\tau x)$ for all x, then $(T\zeta)(x) = (T\zeta)(\tau x)$. (The desired equality is then obtained by induction on n.) Let us prove this. By the definition of T the following equalities hold:

$$(T\zeta)(x) = L\zeta(x) + R \sum_{D \ni \bar{x}} \min_{y \neq \bar{x}, y \in D} \zeta(y),$$

$$(T\zeta)(\tau x) = L\zeta(\tau x) + R \sum_{D \ni \overline{\tau x}} \min_{y \neq \overline{\tau x}, y \in D} \zeta(y).$$

By the hypothesis, $\zeta(\tau x) = \zeta(x)$. Since τ is an automorphism, we have $\bar{\tau} x = \tau \bar{x}$, while the relation $\tau \bar{x} = \overline{\tau x} \in D$ is equivalent to $\bar{x} \in \tau^{-1} D$. Therefore the condition $y \neq \overline{\tau x}, y \in D$ is equivalent to the fact that $z = \tau^{-1} y$ satisfies the condition $z \neq \bar{x}, z \in \tau^{-1} D$. The set of all y such that $y \neq \overline{\tau x}, y \in D$ coincides with the set of all z such that $z \neq \bar{x}, z \in \tau^{-1} D$. Therefore

$$(T\zeta)(\tau x) = L\zeta(x) + R \sum_{\tau^{-1} D \ni \bar{x}} \min_{\substack{z \in \tau^{-1} D \\ z \neq \bar{x}}} \zeta(z).$$

Since τ is an automorphism, the sum over all $\tau^{-1} D$ contains the same terms as the sum over all D. Thus $(T\zeta)(\tau x) = (T\zeta)(x)$, and the proof is completed by induction on n.

THEOREM 2.9. The proof is similar.

THEOREM 3.1. (1) Let us show that if $J(\zeta) = 0$, then any Boolean vector b compatible with ζ defines a satisfying truth assignment.

Indeed, if $J(\zeta) = 0$, i.e., $\sum_{D \in F} \min_{x \in D} \zeta(x) = 0$, then $\min_{x \in D} \zeta x = 0$ for any $D \in F$, i.e., for any clause D there is an $x \in D$ such that $\zeta x = 0$. Since the compatibility of ζ with b means that $\zeta x = 0$ implies that x is true, it follows that every clause D contains a true term, and so D is true.

(2) Now we show that if $J\zeta > 0$, then there exists a Boolean vector b compatible with ζ which is not a satisfying vector for the formula F. Indeed, since $J(\zeta) > 0$, there exists a clause D_0 such that $\min_{x \in D_0} \zeta x > 0$, i.e., $\zeta x > 0$ for all $x \in D_0$. Define a Boolean vector b in the following way: if $\zeta a = 0$, then $ba = 1$; if $\zeta \bar{a} = 0$, then $ba = 0$; if $a \in D_0$, then $ba = 0$; if $\bar{a} \in D_0$, then $ba = 1$; for other a we can define b arbitrarily. As a result, we obtain a Boolean vector for which the clause D_0 is false; hence F is false, i.e., b is compatible with ζ and is not a satisfying collection. The theorem is proved.

THEOREM 3.2. If formula F is satisfiable and b is a satisfying Boolean vector, then for $\zeta x = 1 - bx$ we have $\zeta a + \zeta \bar{a} = 1$ and $J(\zeta) = 0$, whence $j = 0$. In this case, due to Theorem 3.1, all collections that are compatible with those obstacles ζ which minimize J are satisfying collections.

Suppose that the formula F is not satisfiable. Since the set of all $\zeta \geq 0$ for which $\zeta a + \zeta \bar{a} = 1$ is a compact set, and J is a continuous function on this compact, it attains its minimum j at some point ζ. If $j = 0$, then Theorem 3.1 provides a satisfying collection, which contradicts the fact that F is unsatisfiable. Hence $j > 0$. The theorem is proved.

THEOREM 3.3. Evidently

$$\frac{D(\sum A_i)}{D\zeta(a)} = \sum_i \frac{DA_i}{D\zeta(a)};$$

therefore,

$$\frac{DJ}{D\zeta(a)} = \sum_{D \in F} \frac{D}{D\zeta(a)}(\min_{x \in D} \zeta(x)).$$

If $a \notin D$, then $\frac{D}{D\zeta(a)}(\min_{x \in D} \zeta(x)) = 0$. If $a \in D$, then

$$\frac{D}{D\zeta(a)}(\min \zeta x) = \min\{\min_{x \in D\setminus(a)} \zeta x, \infty\} - \min\{\min_{x \in D\setminus(a)} \zeta x, 0\} = \min_{x \in D\setminus(a)} \zeta x,$$

and hence

$$\frac{DJ}{D\zeta(a)} = \sum_{D:D \ni a} \min_{x:x \neq a} \zeta x,$$

which implies the statement of the theorem.

Theorem 3.4 is proved along similar lines.
Theorems 3.5–3.7 are proved similarly to Theorems 3.1–3.3.
Theorem 4.1 is proved by a simple substitution of formula (4.4) in (4.6) for the system of reactions (4.1).
Theorem 4.2 is proved in the text.
Theorem 4.3 is proved exactly as Theorem 1.2.

Appendix 2. The average number of satisfying collections for an arbitrary formula

Suppose that $2k$ literals $x_1, \ldots, x_k, \bar{x}_1, \ldots, \bar{x}_k$ are chosen independently and with equal probability as all entries of a $3 \times d$ matrix. The columns of the matrix are then interpreted as 3-CNF clauses.

This notion of a random formula was introduced by Maslov and Kurierov and used in numerical experiments. In particular, as a result of these experiments the average number of satisfying collections for $d = 30$ and $k = 7$ turned out to be of order 1—3.

Matiyasevich asked whether this empirical result can be justified theoretically. This question is answered by the following result.

THEOREM A.1. *The average number of satisfying collections for a random formula in k variables that contains d clauses is equal to $2^k \left(\frac{7}{8}\right)^d$.*

COROLLARY. *For $k = 7$ and $d = 30$ the average number of satisfying collections is equal to $2^7 \cdot \left(\frac{7}{8}\right)^{30} \approx 2.33$.*

PROOF OF THE THEOREM. The probability that an arbitrary collection of 0's and 1's satisfies a random formula F is (since the columns D_i are independent) equal to the product of these probabilities for individual columns. For each clause we have

$$p(x \vee y \vee z) = 1 - p(\bar{x})p(\bar{y})p(\bar{z}) = 1 - \left(\frac{1}{2}\right)^3 = \frac{7}{8}.$$

Therefore the probability that the formula is satisfied by a given Boolean vector is equal to $\left(\frac{7}{8}\right)^d$, and, since the total number of vectors is 2^k, the average number of satisfying vectors is $2^k \cdot \left(\frac{7}{8}\right)^d$.

Appendix 3. Strong compatibility with satisfiability

Since the conceptual interpretation of the problem has already been described in Section 1, we will limit ourselves to the statement of results. Recall that by Definition 3.1 an obstacle is called *partially regular* if $\zeta a + \zeta \bar{a} > 0$ for any variable a and $\zeta x = 0$ for some literal x.

DEFINITION. A *partial n*-dimensional Boolean vector b is a mapping from some subset of the set of all variables from formula F into $\{0, 1\}$. We say that a partially regular obstacle ζ *defines* a partial Boolean vector b, if $ba = 1 \leftrightarrow \zeta a = 0$, and $ba = 0 \leftrightarrow \zeta \bar{a} = 0$. We say that a partial Boolean vector is a *satisfying* vector for formula F if any clause that contains either a variable for which b is defined or its negation becomes identically true under the substitution of $a =$ "true" (when $ba = 1$), and $a =$ "false" (when $ba = 0$).

This definition easily implies that if a partial Boolean vector b satisfies the satisfiable formula F, then the formula obtained from F by substituting $a =$ "true" (when $ba = 1$) and $a =$ "false" (when $ba = 0$) is also a satisfiable formula.

When the obstacle is regular, these definitions become the conventional ones.

DEFINITION. The iterative method (1.1) is called *strongly compatible* with satisfiability if:
1. for every partially regular obstacle ζ that defines a satisfying partial Boolean vector, $T\zeta$ defines either the same Boolean vector or its extension;
2. if a partial Boolean vector b defined by a partially regular obstacle ζ is not satisfying for F, then $T\zeta$ does not define b.

Since a regular obstacle is a special case of a partially regular obstacle, then a method that is strongly compatible with satisfiability is also compatible with satisfiability.

THEOREM 1.1'. *If an iterative method is strongly compatible with satisfiability and converges to a partially regular obstacle, then the partially Boolean vector determined by this method is a satisfying vector.*

If we know a satisfying partial Boolean vector, then its substitution in F reduces the problem to a problem with fewer variables.

THEOREM 1.2'. *Maslov's iterative method is strongly compatible with satisfiability.*

The proofs are virtually the same as for Theorems 1.1 and 1.2.

Strong compatibility also holds for all the modifications of Maslov's method described in this article.

References

[1] S. Yu. Maslov, *Asymmetry of cognitive mechanisms and its implications*, Semiotics and Information Science No. 20, VINITI, Moscow, 1983, pp. 3–31. (Russian)

[2] R. I. Freidzon, In this volume.
[3] V. Ya. Kreinovich, *Justification of the form of Maslov's operator*, Third All-Union Conf. "Application of Methods of Mathematical Logic", Abstracts of Reports, Tallinn, 1983. (Russian)
[4] E. Ya. Dantsin, In this volume.
[5] I. V. Melichev, In this volume.
[6] Yu. V. Matiyasevich, In this volume.
[7] S. Yu. Maslov and Yu. N. Kurierov, *Strategy for increasing the freedom of choice when recognizing propositional satisfiability*, All-Union Conf. "Methods of Mathematical Logic in Artificial Intelligence and System Programming", Abstracts of Reports, Part 1, Vilnius, 1980, pp. 130–131. (Russian)

Ergodic Properties of Maslov's Iterative Method

M. I. Zakharevich

ABSTRACT. In order to solve the universal search problem of recognizing satisfiability and finding a satisfying truth assignment for formulas of propositional calculus, S. Yu. Maslov suggested a method based on iterating a linear operator constructed from the given formula. Numerical experiments demonstrate that in many cases the problem of satisfiability and finding the satisfying truth assignment can be solved by analyzing the maximum eigenvector of this operator.

In the present paper we prove ergodic theorems on the convergence of the arithmetic mean of iterations in Maslov's method which make it possible to suggest a method of searching for a maximum eigenvector. The proofs make use of only those properties of Maslov's method that are not related to its explicit formula, and so the theorems is this paper can be extended to some generalizations of Maslov's operator

1. Introduction

In order to solve the problem of recognizing satisfiability and searching for a satisfying truth assignment for formulas of the propositional calculus, S. Yu. Maslov suggested a method based on using iterations of an operator in the space $R_+^{(2n)}$ constructed from the given formula. Denote by Λ the set of $2n$ elements interpreted as n logical variables and their negations. Let us realize the cone $R^{(2n)}$ as the space of functions $\xi: \Lambda \to R_+^1$. The operator $T_{R \cdot L}$ which defines Maslov's iterative method is defined by a logical formula written in conjunctive normal form (CNF): $D = \bigwedge_{\alpha \in W} d_\alpha$, where $d_\alpha = a_1^{(\alpha)} \vee \ldots \vee a_{k_\alpha}^{(\alpha)}$, $a_i^{(\alpha)} \in \Lambda$, where W is a finite set,

$$(T_{R \cdot L} \xi)(a) = R \cdot \xi(a) + L \cdot \sum_{\substack{d \in \bar{a} \\ b \neq \bar{a}}} \min_{\substack{b \in D \\ b \neq \bar{a}}} \xi(b),$$

where $R, L \geq 0$.

Maslov investigated an important case of 2-CNF-formulas, i.e., those formulas that contain at most two literals in every clause d_α (in this case the operator $T_{R \cdot L}$ is a linear one) and proved the following theorem.

THEOREM [4]. *Let $R \cdot L > 0$.*

(1) For any $\xi \in \text{int } R_+^{2n}$ (i.e., $\xi_i > 0$ for all i) we have
$$\frac{T_{R,L}^N \xi}{\|T_{R,L}^N \xi\|} \to \bar{\xi},$$
where $\bar{\xi}$ is a maximum eigenvector of the operator $T_{R,L}$.

(2) If a is a literal such that the corresponding coordinate $\bar{\xi}(a)$ has the maximum value among the coordinates of the vector $\bar{\xi}$, then, setting $a =$ "true" and $\bar{a} =$ "false" in D we obtain the formula D' in a smaller number of variables, which is satisfiable if and only if the original formula is.

A theorem proved by Maslov gives the following tactic for searching a satisfying truth assignment for 2-CNF-formulas. Starting from a vector $\xi \in \text{int } R_+^{(2n)}$ (for example, $\xi_0(a) = 1$), perform several (N) iterations of the operator $T_{R,L}$. Let $\xi_N = T_{R,L}^N \xi_0$. Find a literal a for which $\xi_N(a)$ attains its maximum. Set a in formula D to "false" and $\bar{a}$ to "true", and repeat the above procedure for the resulting formula having a smaller number of variables. By virtue of Maslov's theorem, for a sufficiently large N we either arrive at a satisfying truth assignment, or obtain a proof that the formula is unsatisfiable.

In the general case (not just 2-CNF) none of the statements of Maslov's theorem holds. Nevertheless, as was shown in numerical experiments, part (2) of this theorem is often true. However, there is no convergence $T^N \xi / \|T^N \xi\| \to \xi$ (this was also demonstrated in numerical experiments). In this paper we prove ergodic theorems on the convergence of the arithmetic mean of iterations of the mapping $T_{R,L}$. These theorems make it possible to suggest an iterative method of searching for a maximum eigenvector generalizing Maslov's tactics. The proofs make use of only those properties of Maslov's method that are not related to its formula, and so the theorems is this paper can be extended to some generalizations of Maslov's operator. Nowhere below shall we use the explicit form of Maslov's operator. Maslov's operators $T_{R,L}$ belong to the class of operators studied in the mathematical theory of economical dynamics and define a subclass of von Neumann models (see [1,2]). The proof techniques used in this article are based on certain facts from the theory of models in economical dynamics, and ergodic theory for linear operators in Banach spaces, as well as on papers [5,6].

The main theorem is Theorem 3. In a much more general form for stochastic multivalued mappings related to the models of economical dynamics it was announced in [6] (the complete proof is contained in the author's dissertation). In this paper we publish the proof of the deterministic version of the theorem from [6]. The main results of the present paper were announced in [7]. The author is grateful to V. Ya. Kreinovich, R. I. Freidzon, and I. V. Melichev for their help and discussions in the course of the work.

2. Main definitions. Properties of Maslov's operators

DEFINITION. (1) A mapping $T: R_+^m \to R_+^m$ is said to be *piecewise linear* if there exists a partition of R_+^m into a finite number of polyhedral domains such that the restriction of T to each domain is linear.

(2) A mapping T is said to be *homogeneous* (to be more precise, *positive homogeneous*) if $T(\lambda \xi) = \lambda T(\xi)$ for any ξ and $\lambda > 0$.

(3) A mapping T is said to be *concave* if
$$T(\alpha\xi_1 + (1-\alpha)\xi_2) \geq \alpha T(\xi_1) + (1-\alpha)T(\xi_1) + (1-\alpha)T(\xi_2)$$
for any $\xi_1, \xi_2, \alpha \in (0,1)$. (In this article inequalities between vectors should be understood coordinatewise, i.e., $\xi \geq \eta$ means that $\xi_i \geq \eta_i$ for all i.)

In what follows we consider only norms on R^m for which $\xi_1 \leq \xi_2$ implies that $\|\xi_1\| \leq \|\xi_2\|$.

(4) A homogeneous concave mapping $T: R_+^m \to R_+^m$ is said to be *superlinear*.

(5) A mapping T is said to be *monotone* if for any $\xi_1 \geq \xi_2$ in R_+^m we have $T(\xi_1) \geq T(\xi_2)$.

(6) A vector $\xi \in R_+^m$ ($\xi \neq 0$) is called an *eigenvector* of the mapping T if $T(\xi) = \lambda\xi$, where λ is a nonnegative number called the *eigenvalue* of T. An eigenvector ξ is said to be *maximal* if λ takes the maximum possible value.

(7) An *invariant face* for the mapping $T: R_+^m \to R_+^m$ is a set $\Gamma_I = \{\xi \in R_+^m : \xi_i = 0, i \in I\}$ (where $I = \{1, 2, \ldots, m\}$) such that $T(\Gamma_I) \subset \Gamma_I$.

(8) A linear functional on R_m taking nonnegative values on the cone G_+^m will be called a *nonnegative functional*. The set of such functionals forms a cone $(R_+^m)^*$ isomorphic to the cone R_+^m.

In this article we deal mainly with continuous superlinear mappings and monotone mappings $T: R_+^m \to R_+^m$. Such mappings are a special case of more general multivalued mappings that define models of von Neumann-Gale economical dynamics (see [1]).

We now list some properties of superlinear mappings proved in the theory of models of economical dynamics. The main source on this theory is the book [1].

THEOREM ([1, 2]). (1) *If $T: R_+^m \to R_+^m$ is a superlinear mapping, then T is a monotone mapping.*

(2) *For any monotone homogeneous mapping $T: R_+^m \to R_+^m$ there exists a maximal eigenvector.*

(3) *Let T be a superlinear piecewise linear mapping. Then there exist a vector $\hat{\xi} \in R_+^m$ and a functional $\hat{p} \in (R_+^m)^*$ such that $T(\hat{\xi}) \geq \lambda\hat{\xi}$; $\hat{p}(T(\xi)) \leq \lambda\hat{p}(\xi)$ for any $\xi \in R_+^m$; $\hat{p}(\hat{\xi}) > 0$, where λ is the maximal eigenvalue of the mapping T. Such a triple $(\hat{\xi}, \hat{p}, \lambda)$ is a called a von Neumann state of equilibrium.*

(4) (*see* [2]). *Let T be a piecewise linear superlinear mapping. Then for any $\xi \in R_+^m$ there exists $k \leq m - 1$ such that*
$$\varlimsup_{n \to \infty} \frac{\|T^n(\xi)\|}{\lambda^n n^k} < +\infty.$$

3. Ergodic theorems

In this section we prove ergodic theorems for superlinear mappings.

THEOREM 2. *Let T be a superlinear piecewise linear mapping. Then*
$$\lim_{n \to \infty} \|T^n \xi\|^{\frac{1}{n}} = \lambda$$
for any $\xi \in \operatorname{int} R_+^m$, where λ is the maximal eigenvalue. The convergence is uniform on every compact set lying in $\operatorname{int} R_+^m$.

PROOF. Let $\bar{\xi}$ be the maximal eigenvector, i.e., $T\bar{\xi} = \lambda\bar{\xi}$. Take $\xi \in \operatorname{int} R_+^m$. There exists a constant $c > 0$ such that $c\bar{\xi} \leq \xi$. Hence it follows that $T^n\xi \geq cT^n\bar{\xi} = c\lambda^n\bar{\xi}$. Since the norm is monotone, we have $\lim \|T^n\xi\|^{\frac{1}{n}} \geq \lambda$. On the other hand, by virtue of part (4) of Theorem 1, there exists an integer k such that $\overline{\lim} \|T^n\xi\|/\lambda^n n^k < +\infty$. This implies that $\overline{\lim} \|T^n\xi\| \leq \lambda$. By comparing the two inequalities we obtain the convergence to λ. Its uniformity follows from the fact that any two points of the compact lying in the interior of R_+^m can be estimated in terms of each other, i.e., if $K \subset \operatorname{int} R_+^m$, then there exist c_1, c_2 such that for any $\xi_1, \xi_2 \in K$ we have $c_1\xi_1 \leq \xi_2 \leq c_2\xi_1$. The theorem is proved.

Let us introduce the following assumption about the mapping T:

(*) There exist a vector $\bar{\xi} \in \operatorname{int} R_+^m$ and a number $\lambda > 0$ such that $T\bar{\xi} = \lambda\bar{\xi}$.

The following Theorems 3,4,5 will be proved under this assumption. The general case will be considered later.

We shall consider the case of monotone uniform mappings.

THEOREM 3. *Let T be a monotone homogenous mapping satisfying the condition* (*). *Then*

$$\lim_{N\to\infty} \frac{1}{N} \sum_{n=0}^{N} \frac{T^n\xi}{\lambda^n} \stackrel{\text{def}}{=} P(\xi)$$

for any $\xi \in \operatorname{int} R_+^m$. The convergence is uniform on compacts lying in $\operatorname{int} R_+^m$; P is a homogeneous monotone mapping $\operatorname{int} R_+^m \to \operatorname{int} R_+^m$ such that $P(T\xi) = P(\xi)$.

REMARK. If T is a superlinear mapping, then for $P(\xi)$ the inequality $T(P(\xi)) \geq \lambda P(\xi)$ holds, and, therefore, $T^n(P(\xi))/\lambda^n \to \tilde{\xi}$ as $n \to \infty$; $T\tilde{\xi} = \lambda\tilde{\xi}$.

PROOF OF THE REMARK. The fact that the mapping is concave implies that

$$T\left(\frac{1}{N}\sum_{n=0}^{N}\frac{T^n\xi}{\lambda^n}\right) \geq \frac{1}{N}\sum_{n=0}^{N}\frac{T^{n+1}\xi}{\lambda^n} = \lambda\frac{1}{N}\sum_{n=0}^{N}\frac{T^{n+1}\xi}{\lambda^{n+1}}$$

$$= \lambda\left(\frac{1}{N}\sum_{n=0}^{N}\frac{T^n\xi}{\lambda^n} + \frac{1}{N}\left(\frac{T^{N+1}\xi}{\lambda^{N+1}} - \xi\right)\right) \underset{N\to\infty}{\to} \lambda P(\xi).$$

The fact we have just used, viz., that the sequence $\{\|T^N\xi\|/\lambda^N\}$ is bounded, will be demonstrated in the proof of Theorem 3. The remark is proved.

Before we begin with the proof of Theorem 3 let us first prove several auxiliary statements.

Denote by X the space of Borel homogeneous functions on R_+^m bounded on the intersection of the unit ball with R_+^m. Introduce a norm in X by setting $\|f\| = \sup_{\|\xi\|=1}|f(\xi)|$. Denote by X_+ the subspace of X consisting of monotone functions.

LEMMA 1. *For functions in X_+ the following properties are satisfied.*

(1) *If $f \in X_+$, then f is continuous on $\operatorname{int} R_+^m$.*

(2) *Suppose that the sequence $\{f_n\} \subset X_+$ satisfies $\sup_n \|f_n\| < +\infty$. Then there exist a subsequence of indices $\{n_k\}$ and $f \in X$ such that $\lim_{k\to\infty} f_{n_k} = \bar{f}(\xi)$ for any $\xi \in \operatorname{int} R_+^m$.*

PROOF OF THE LEMMA. (1) Let $\{\xi_n\} \subset \text{int}\, R_+^m$, $\xi_n \to \xi$, $\xi \in \text{int}\, R_+^m$. Since all the points ξ' sufficiently close to ξ satisfy the inequality $(1-\varepsilon)\xi' \leq \xi \leq (1+\varepsilon)\xi'$, then for any $\varepsilon > 0$ there exists N such that $(1-\varepsilon)\xi_n \leq \xi \leq (1+\varepsilon)\xi_n$ for $n > N$. Since $f \in X_+$ is monotone and homogeneous, we have $(1-\varepsilon)f(\xi_n) \leq \xi \leq (1+\varepsilon)f(\xi_n)$. Hence, by going to the limit, we obtain $(1-\varepsilon)\overline{\lim} f(\xi_n) \leq f(\xi) \leq (1+\varepsilon)\underline{\lim}(\xi_n)$. When $\varepsilon \to 0$ we have $\lim f(\xi_n) = f(\xi)$, i.e., f is continuous on $\text{int}\, R_+^m$.

(2) Let E be the set of all vectors in R_+^m having rational coordinates. Using the diagonal procedure and the fact that the sequences $\{f_n\}$ are bounded on the unit ball, we can select a subsequence $\{f_{n_k}\}$ such that $\lim_{k\to\infty} f_{n_k}(\xi) \stackrel{\text{def}}{=} f(\xi)$ for any $\xi \in E$. This defines a function $\bar{f}\colon E \to R_+^1$. The argument in the proof of part (1) demonstrates that the function $\bar{f}$ is uniformly continuous on the set $E \cap B_1$. Therefore, the function $\bar{f}\colon E \to R_+^1$ can be continued to a continuous function $\bar{f}$ on $\xi \in \text{int}\, R_+^m$. Indeed, let $\{\xi_k\} \subset E$ be such that $\xi_k \to \xi$. For any $\varepsilon > 0$ there is $N(\varepsilon)$ such that $(1-\varepsilon)\xi_k \leq \xi \leq (1+\varepsilon)\xi_k$ for $k > N(\varepsilon)$. Hence it follows that

$$\overline{\lim_{t\to\infty}} f_{n_i}(\xi) \leq (1+\varepsilon)\lim_{i\to\infty} f_{n_i}(\xi_k) = (1+\varepsilon)\bar{f}(\xi_k) \leq \frac{1+\varepsilon}{1-\varepsilon}\bar{f}(\xi_k),$$

$$\underline{\lim_{t\to\infty}} f_{n_i}(\xi) \leq (1-\varepsilon)\lim_{i\to\infty} f_{n_i}(\xi_k) = (1-\varepsilon)\bar{f}(\xi_k) \leq \frac{1-\varepsilon}{1+\varepsilon}\bar{f}(\xi_k).$$

It follows that $\lim_{t\to\infty} f_{n_i}(\xi) = \bar{f}(\xi)$. Lemma 1 is proved.

We now state a result on the convergence of averages of powers of a linear operator in a Banach space. It is proved, for example, in [3].

LEMMA 2. *Let X be a Banach space and A a bounded linear operator on it. Suppose that the following conditions are satisfied:*
(a) $\sup_n \|A^n\| < +\infty$.
(b) *There exists a set $X' \subset X$ whose linear hull is dense everywhere such that for each $f \in X'$ there exists $\bar{f} \in X$ which is a weak limit point for the sequence*

$$A_n f = \frac{1}{n} \sum_{k=0}^{n} A^k f.$$

Then $\lim_{n\to\infty} A_n f = \bar{f}$, where the limit is taken with respect to the norm in the space X.

COROLLARY. *Let X be a Banach space, A a bounded linear operator on it, $f \in X$. Suppose that*
(a) $\sup_n \|A^n\| < +\infty$, *and*
(b') *there exists $\bar{f} \in X$ such that $\bar{f}$ is a weak limit point of the sequence $\{A_n f\}$. Then $\lim_{n\to\infty} A_n f = \bar{f}$ with respect to the norm X.*

PROOF. Consider the set $X' = \{f \in X\colon f \text{ satisfies condition } b'\}$. It is not hard to show that $AX' \subset X'$, i.e., that X' is invariant under A. Denote by X_1 the closure of the linear hull of the set X'. Then X_1 is an invariant subspace for A. Here $f \in X_1$ and $\bar{f} \in X_1$. To complete the proof it remains to apply Lemma 2 with the role of the space X played by X_1. The corollary is proved.

PROOF OF THEOREM 3. By replacing the mapping T with $\frac{T}{\lambda}$ we can assume that $\lambda = 1$. Thus condition (*) is reduced to the existence of $\bar{\xi} \in \text{int } R_+^m$ such that $T\bar{\xi} = \bar{\xi}$.

We shall prove that $\sup_{n, \|\xi\|=1} \|T^n \xi\| < +\infty$. Indeed, the fact that $\bar{\xi} \in \text{int } R_+^m$ implies that there exists $c > 0$ such that $\xi \leq c\bar{\xi}$ for any $\xi \in R_+^m$ such that $\|\xi\| = 1$. Hence it follows that $T^n \xi \leq cT^n \bar{\xi} = c\bar{\xi}$ and $\|T^n \xi\| \leq c\|\bar{\xi}\| = \text{const}$. The statement that the powers are bounded is proved. Denote

$$P_N(\xi) = \frac{1}{N} \sum_{n=0}^{N} T^n \xi.$$

The above implies that

$$\|P_n(\xi)\| \leq \frac{1}{N} \sum_{n=0}^{N} \|T^n \xi\| \leq \text{const}.$$

In order to prove the theorem it is sufficient to demonstrate that for any $i, 1 \leq i \leq m$, the sequence $\{f(P_N(\xi))\}$, where $f(\xi) = \xi_i$, converges uniformly on any compact set $K \subset \text{int } R_+^m$. Since F is linear, we have

$$f(P_N(\xi)) = \frac{1}{N} \sum_{n=0}^{n} f(T^n \xi).$$

The last expression has a limit not just for linear functions but for any function $f \in X_+$. Let K be a compact, $K \in \text{int } R_+^m$, $f_0 \in X_+$. It follows from condition (*) that there exist $c_1, c_2 > 0$ such that $c_1 \bar{\xi} \leq \xi \leq c_2 \bar{\xi}$ for any $\xi \in K$. Denote by L the subset of R_+^m generated by the set $\{\xi : c_1 \bar{\xi} \leq \xi \leq c_2 \bar{\xi}\}$. Since L is monotone, the operator T is invariant, i.e., $TL \subset L$. Define the operator A by the formula

$$Af(\xi) = f(T\xi) \quad \text{for} f \in X.$$

Introduce the function $\varphi \in X$ as follows:

$$\varphi(\xi) = \begin{cases} f_0(\xi) & \text{for } \xi \in L, \\ 0 & \text{for } \xi \notin L. \end{cases}$$

The definition of the operator A implies that

$$A^n \varphi(\xi) = \begin{cases} f_0(T^n \xi) & \text{for } \xi \in L, \\ 0 & \text{for } \xi \notin L. \end{cases}$$

Let us prove that the operator A acts in the space X and the function $\varphi \in X$ satisfies the conditions of Lemma 2.

Condition (a) follows from the fact that powers of the mapping T are uniformly bounded. The cone X_+ is invariant under the operator A $(AX_+ \subset X_+)$. Indeed, since the mapping T is monotone, we see that $Af(\xi) = f(T\xi)$ is a monotone function if $f \in X_+$. The cone X_+ is convex, and so $\frac{1}{N} \sum_{n=0}^{N} f_0(T^n \xi) \in X_+$ for all

N. Making use of Lemma 1, we see that there exists a function $\bar{f} \in X$ such that there is a subsequence for which

$$\lim_{k \to \infty} \sum_{n=0}^{N_k} f_0(T^n \xi) = \bar{f}(\xi)$$

for any $\xi \in \operatorname{int} R_+^m$. Introduce the function

$$\bar{\varphi}(\xi) = \begin{cases} \bar{f}(\xi) & \text{for } \xi \in L, \\ 0 & \text{for } \xi \notin L. \end{cases}$$

By the definition of the functions φ and $\bar{\varphi}$ we have

$$\frac{1}{N_k} \sum_{n=0}^{N_k} A^n \varphi(\xi) \to \bar{\varphi}(\xi).$$

Consider the space $X(L)$ of continuous homogeneous functions on L with the norm $\|f\|_{X(L)} = \sup_{\|\xi\|=1} |f(\xi)|$, which can be regarded as a closed subspace of the space X if $f \in X(L)$ is continued by zero to the entire space R_+^m. Note that $\varphi_N, \bar{\varphi} \in X(L)$, where $\varphi_N = \frac{1}{N} \sum_{n=0}^{N} A^n \varphi$. This follows from Lemma 1. We have shown that $\varphi_{N_k} \to \bar{\varphi}$ pointwise, i.e., in the weak topology of the space $X(L)$. An application of Lemma 2 yields

$$\lim_{N \to \infty} \frac{1}{N} \sum_{n=0}^{N} A^n \varphi(\xi) = \lim_{N \to \infty} \frac{1}{N} \sum_{n=0}^{N} f_0(T^n \xi) = \bar{\varphi}(\xi)$$

uniformly with respect to $\xi \in L$; $\|\xi\| = 1$. Thus, we have proved the existence of the limit

$$\lim_{N \to \infty} \frac{1}{N} \sum_{n=0}^{N} T^n \xi \stackrel{\text{def}}{=} P(\xi)$$

uniformly with respect to $\xi \in K$. We have

$$P(T\xi) = \lim_{N \to \infty} \frac{1}{N} \sum_{n=0}^{N} T^{n+1} \xi = \lim_{N \to \infty} \frac{1}{N} \sum_{n=1}^{N+1} T^n \xi$$

$$= \lim_{N \to \infty} \frac{1}{N} \sum_{n=0}^{N} T^n \xi + \lim_{N \to \infty} \frac{1}{N}(T^{N+1}\xi - \xi) = P(\xi).$$

The proof of Theorem 3 is complete.

THEOREM 4. *Let T be a superlinear, piecewise linear mapping satisfying condition (∗). Suppose that the sequence $\{M(N)\}$, where $M(N)$ is a positive integer, satisfies the condition $\lim_{N \to \infty} \frac{M(N)}{N} = +\infty$. Let $\xi \in \operatorname{int} R_+^m$. Denote*

$$\lambda_N = \lambda_N(\xi) = \|T^{M(N)}\xi\|^{1/M(N)}.$$

Then

$$\lim_{N \to \infty} \sum_{n=0}^{N} \frac{T^n \xi}{\lambda_N^n} = P(\xi),$$

where $P(\xi)$ is defined in Theorem 3. The convergence is uniform on any compact set lying in $\operatorname{int} R^m_+$.

REMARK. The meaning of this theorem is that the number λ which is equal to the limit of $\|T^N\xi\|^{\frac{1}{N}}$ as $N \to \infty$ is replaced by the expression under the limit. Thus, in order to find the limit $P(\xi)$, only one limit transition is required (instead of two).

PROOF OF THEOREM 4. Let $\bar\xi$ be the maximal eigenvector, $\|\bar\xi\| = 1$. In view of condition $(*)$, for any compact $K \subset \operatorname{int} R^m_+$, there exist $c_1, c_2 > 0$ such that $c_1 \bar\xi \le \xi \le c_2 \bar\xi$ for any $\xi \in K$. Hence it follows that

$$c_1 \lambda^{M(N)} \le \|T^{M(N)}\xi\| \le c_2 \lambda^{M(N)},$$
$$c_1^{\frac{1}{M(N)}} \lambda \le \lambda_N \le c_2^{\frac{1}{M(N)}} \cdot \lambda.$$

Using these inequalities we obtain

$$c_2^{-\frac{N}{M(N)}} \cdot \frac{1}{N} \sum_{n=0}^{N} \frac{T^N \xi}{\lambda^n} \le \frac{1}{N} \sum_{n=0}^{N} \frac{T^N \xi}{\lambda_N^n} \le c_1^{-\frac{N}{M(N)}} \cdot \frac{1}{N} \sum_{n=0}^{N} \frac{T^n \xi}{\lambda^n}.$$

Hence the statement of Theorem 4.

Denote $\bar T(\xi) = T\xi/\|T\xi\|$.

THEOREM 5. Let T be as in Theorem 4. Suppose that $Q\colon N \to N$ is a sequence of positive integers satisfying the condition $Q(N) \to \infty$ as $n \to \infty$. Then for any $\xi \in \operatorname{int} R^m_+$ there exists the limit

$$\lim_{N \to \infty} \bar T^{Q(N)} \left(\frac{1}{N} \sum_{n=0}^{N} \frac{T^n \xi}{\lambda_N^n} \right) \stackrel{\text{def}}{=} \tilde\xi,$$

where $\tilde\xi$ is a maximal eigenvector T corresponding to the eigenvalue λ: $T\tilde\xi = \lambda\tilde\xi$. The convergence is uniform on any compact lying in the interior of R^m_+.

REMARK. This theorem gives an iterative procedure for finding the maximal eigenvector in the case $(*)$.

PROOF OF THEOREM 5. In Theorem 3 it was proved that $P(\xi) \in \operatorname{int} R^m_+$. By Theorem 4

$$4\frac{1}{N} \sum_{n=0}^{N} \frac{T^n \xi}{\lambda_N^n} \to P(\xi).$$

Therefore, for any $\varepsilon > 0$ we have

$$(1-\varepsilon)P(\zeta) \le \frac{1}{N} \sum_{n=0}^{N} \frac{T^n \zeta}{\lambda_N^n} \le (1+\varepsilon)P(\xi)$$

for any $N > N(\varepsilon)$. Hence it follows that

$$(1-\varepsilon)T^{Q(N)}P(\zeta) \le T^{Q(N)}\left(\frac{1}{N}\sum_{n=0}^{N} \frac{T^n \zeta}{\lambda_N^n}\right) \le (1+\varepsilon)T^{Q(N)}P(\xi)$$

for any $N > N(\varepsilon)$. Thus it is sufficient to prove that
$$\tilde{T}^{Q(N)}P(\xi) \underset{N\to\infty}{\to} \tilde{\xi}, \quad T\tilde{\xi} = \lambda\tilde{\xi}.$$

It follows from the remark to Theorem 3 that
$$\frac{T^{(Q(N))}}{\lambda^{Q(N)}} \uparrow \tilde{\xi}_1 \neq 0, \quad T\tilde{\xi}_1 = \lambda\tilde{\xi}_1.$$

Therefore
$$\left\|\frac{T^{Q(N)}P(\xi)}{\lambda^{Q(N)}}\right\| \to \|\tilde{\xi}_1\|,$$
$$\tilde{T}^{Q(N)}P(\xi) = \frac{T^{Q(N)}P(\xi)}{\|T^{Q(N)}P(\xi)\|} \to \frac{\tilde{\xi}_1}{\|\tilde{\xi}_1\|} = \tilde{\xi}.$$

Theorem 5 is thus proved.

Now consider mappings T that do not necessarily satisfy condition $(*)$.

THEOREM 6. *Let $T: R_+^m \to R_+^m$ be a superlinear piecewise linear mapping. Suppose that the maximal eigenvector satisfies the condition $\lambda > 0$. Then one of the following three statements holds.*

(1) *The mapping T has a nondegenerate eigenvector (which is equivalent to condition $(*)$).*

(2) *There exists a proper face Γ of the cone R_+^m invariant under T such that for any $\xi \in \operatorname{int} R_+^m$ we have*
$$\lim_{N\to\infty} \operatorname{dist}\left(\frac{T^N\xi}{\|T^N\xi\|}, \Gamma\right) = 0.$$

The face Γ contains the maximal eigenvector.

(3) *There exists a proper face Γ of the cone R_+^m invariant under T such that for any $\xi \in \operatorname{int} R_+^m$ we have*
$$\lim_{N\to\infty} \operatorname{dist}\left(\frac{\sum_{n=0}^N \frac{T^N\xi}{\lambda_N^n}}{\left\|\sum_{n=0}^N \frac{T^N\xi}{\lambda_N^n}\right\|}, \Gamma\right) = 0,$$

where $\lambda_N = \|T^{M(N)}\xi\|^{\frac{1}{M(N)}}$ and $M(N)$ is an arbitrary sequence of positive integers such that $M(N) \to +\infty$ as $N \to +\infty$. The face Γ contains the maximal eigenvector.

REMARK. Maslov's operator $T_{R,L}$ satisfies all the conditions of Theorem 6 if $R, L > 0$.

PROOF OF THEOREM 6. Consider the triple $(\hat{\xi}, \hat{\pi}, \lambda)$ which is the von Neumann equilibrium state for the mapping T. Recall that this triple satisfies the conditions $T\hat{\xi} \geq \lambda\hat{\xi}$; $\hat{p}(T\xi) \leq \lambda\hat{p}(\xi)$ for any ξ; $\hat{p}(\hat{\xi}) > 0$; λ a maximal eigenvalue. Suppose that condition $(*)$ is not satisfied. Consider two cases, (a) and (b).

(a) There exists a von Neumann equilibrium state $(\hat{\xi}, \hat{p}, \lambda)$ such that $\hat{\xi} \in \operatorname{int} R_+^m$. We shall prove that in this case condition (2) of Theorem 6 holds. Consider the sequence of vectors $\tilde{\xi}_N = \frac{T^N \hat{\xi}}{\lambda^N}$. By the definition of the equilibrium state we have $\tilde{\xi}_{N+1} \geq \tilde{\xi}_N \geq \hat{\xi}$.

Let us prove that $\{\tilde{\xi}_N\}$ is not bounded. Suppose that the sequence $\{\tilde{\xi}_N\}$ is bounded. Then the fact that it is monotone implies that there exists a finite limit $\xi = \lim_{N \to \infty} \tilde{\xi}^N = \lim_{N \to \infty} T^n \hat{\xi}/\lambda^N$.

The condition of continuity implies that $T\tilde{\xi} = \lim_{N \to \infty} T^{N+1}\xi/\lambda^N = \lambda\tilde{\xi}$. Since $\tilde{\xi} \geq \xi \in \operatorname{int} R_+^m$, then $\tilde{\xi} \in \operatorname{int} R_+^m$. The last statement contradicts the assumption that condition (∗) is not satisfied. Therefore, $\lim_{N \to \infty} \|T^N \hat{\xi}\|/\lambda^N = \infty$.

Consider an arbitrary vector $\xi \in \operatorname{int} R_+^m$. There exists $c > 0$ such that $\xi \geq c\hat{\xi}$. Hence it follows that

$$\lim_{N \to \infty} \frac{\|T^N \xi\|}{\lambda^n} \geq c \lim_{N \to \infty} \frac{\|T^N \hat{\xi}\|}{\lambda^N} = +\infty.$$

By definition, $\hat{p}(T^N \hat{\xi}) = \lambda^N \hat{p}(\hat{\xi})$. Hence it follows that $\hat{p} \notin \operatorname{int}(R_+^m)^*$, i.e., $\hat{p}(\xi) = 0$ for some $\xi \neq 0$, as otherwise the sequence $\{T^N \hat{\xi}/\lambda^N\}$ would have been bounded, which is impossible. For any $\xi \in \operatorname{int} R_+^M$ we have

$$\lim_{N \to \infty} \hat{p}\left(\frac{T^N \xi}{\|T^N \xi\|}\right) \leq \lim_{N \to \infty} \frac{\lambda^N \hat{p}(\xi)}{\|T^N \xi\|} = 0.$$

Thus all limit points of the sequence $\{T^N \xi/\|T^N \xi\|\}$ belong to the face $\Gamma = \operatorname{Ker} \hat{p} \cap R_+^m$. Since $\hat{p}$ is degenerate, statement (2) of Theorem 6 is proved. The definition of $\hat{p}$ implies that the face Γ is invariant.

Now consider the next case.

(b) By the definition of the equilibrium state $(\hat{\xi}, \hat{p}, \lambda)$ we have $\hat{\xi} \notin \operatorname{int} R_+^m$. Denote by P the set of all vectors $\tilde{\xi} \in R_+^m$ such that $T\tilde{\xi} \geq \lambda\tilde{\xi}$. The fact that the mapping is monotone implies that $TP \subset P$. Denote by Γ the smallest face of the cone R_+^m containing the set P. The definition implies that the face Γ is invariant. Let us prove that $\Gamma \neq R_+^m$. Since T is superlinear, the set P is a closed cone, i.e., $\alpha_1 \xi_1 + \alpha_2 \xi_2 \in P$ for any $\xi_1, \xi_2 \in P$, $\alpha_1, \alpha_2 \geq 0$.

Hence it follows that $\Gamma = \{\xi \in R_+^m : \exists \tilde{\xi} \in P : \tilde{\xi} \geq \xi\}$. Thus if $\Gamma = R_+^m$, then $\operatorname{int} R_+^m \cap P \neq 0$, which contradicts condition (b). Therefore, $\Gamma \neq R_+^m$.

Let $\xi \in \operatorname{int} R_+^m$. Denote

$$\xi_N = \sum_{n=0}^{N} \frac{T^n \xi}{\lambda_N^n}.$$

Let us prove that

$$\lim_{N \to \infty} \operatorname{dist}\left(\frac{\xi_N}{\|\xi_N\|}, \Gamma\right) = 0.$$

We shall first prove that $\underline{\lim}_{N \to \infty} \|\xi_N\| = \infty$. Indeed, let $\bar{\xi}$ be the maximal eigenvector. Then, since $\xi \in \operatorname{int} R_+^m$, there exists $c > 0$ such that $\xi \geq c\bar{\xi}$. Making use of the fact that both the mapping T and the norm are monotone, we obtain

$$\|\xi_N\| = \left\|\sum_{n=0}^{N} \frac{T^n \xi}{\lambda_N^n}\right\| \geq \left\|\sum_{n=0}^{N} \frac{T^n \bar{\xi}}{\lambda_N^n}\right\| = \|\bar{\xi}\| \sum_{n=0}^{N} \frac{\lambda^n}{\lambda_N^n}.$$

Since by Theorem 2 $\lim_{N\to\infty} \lambda_N = \lambda$, then for any $\varepsilon > 0$ for all sufficiently large N we have $\lambda > (1-\varepsilon)\lambda_N$. Therefore, for all such N we have

$$\lim_{N\to\infty} \sum_{n=0}^{N} \frac{\lambda^n}{\lambda_N^n} \geq \lim_{N\to\infty} \sum_{n=0}^{N} (1-\varepsilon)^n = \frac{1}{\varepsilon}.$$

Hence $\underline{\lim}_{N\to\infty} \|\xi_N\| > \frac{1}{\varepsilon}$ for any ε, i.e., $\underline{\lim}_{N\to\infty} = +\infty$.

Let $\tilde{\xi}$ be an arbitrary limit point of the sequence $\{\xi_N/\|\xi_N\|\}$, i.e.,

$$\tilde{\xi} = \lim_{i\to\infty} \frac{\xi_{N_i}}{\|\xi_{N_i}\|}.$$

Now we have

$$T\tilde{\xi} = \lim_{i\to\infty} T\left(\sum_{n=0}^{N_i} \frac{T^n \xi}{\lambda_{N_i}^n}\right) \geq \lim_{i\to\infty} \frac{\sum_{n=0}^{N_i} \frac{T^{n+1}\xi}{\lambda_{N_i}^n}}{\|\xi_{N_i}\|} \quad \text{(by superlinearity)}$$

$$= \lim_{i\to\infty} \left(\frac{\sum_{n=0}^{N_i} \frac{T^{n+1}\xi}{\lambda_{N_i}^n}}{\|\xi_{N_i}\|} \cdot \lambda_{N_i}\right) + \lim_{i\to\infty} \left(\frac{T^{N_i+1}\xi}{\lambda_{N_i}^{N_i}\|\xi_{N_i}\|} - \frac{\xi}{\|\xi_{N_i}\|}\right)$$

$$= \lambda\tilde{\xi} + \lim_{i=0} \left(\frac{T^{N_i+1}\xi}{\|T^{N_i}\xi\|\|\xi_{N_i}\|} - \frac{\xi}{\|\xi_{N_i}\|}\right) = \lambda\tilde{\xi},$$

since $\lim_{i=\infty} \|\xi_{N_i}\| = \infty$ and

$$\frac{\|T^{N_i+1}\xi\|}{\|T_\xi^{N_i}\|} \leq \sup_\eta \frac{\|T\eta\|}{\|\eta\|} = \sup_{\|\eta\|=1} \|T\eta\| < +\infty.$$

Consequently, $\tilde{\xi} \in P \subset \Gamma$. Theorem 6 is proved completely.

4. Algorithm solving the satisfiability problem

Our theorems make it possible to suggest the following heuristic algorithm for searching for the maximal eigenvector of Maslov's operator. The algorithm generalizes Maslov's tactic for searching for a satisfiable set for 2-CNF-formulas. The algorithm includes two stages. At the first stage one searches for a face Γ invariant under Maslov's operator $T = T_{R,L}$ containing the maximal eigenvector. The second stage consists in applying an iterative process in searching for the maximal eigenvector in this face Γ. Let us describe the algorithm step by step:

Choose $N \gg 1$, $Q \gg 1$, $M \gg N$, $0 < \varepsilon \ll 1$.

Stage 1.

Step 1. Set $i = 1$, $\Gamma_1 = R_+^m$.

Step 2. If $\Gamma_j = \{\eta: \eta_k = 0 \text{ for } k \in I\}$, then set

$$(\xi_0(j))_k = \begin{cases} 0, & \text{if } \eta_k = 0, \\ 1, & \text{if } \eta_k \neq 0. \end{cases}$$

Step 3. Run N iterations of the operator $T^{R,L}$ starting with the vector $\xi_0 = \xi_0(j)$. Compute $\xi_N = T_{R,L}^N \xi_0 / \|T_{R,L}^N \xi_0\|$. Define the face $\Gamma_{j+1} = \{\eta \in \Gamma_j: \eta_i = 0 \text{ if } (\xi_N)_i < \varepsilon\}$. If $\Gamma_j \neq \Gamma_{j+1}$, then go to Step 4, otherwise go to Step 5.

Step 4. (Check that Γ_{j+1} is invariant.) Apply $T_{R,L}$ to the vector $\xi_0(j+1)$. If $T_{R,L}(\xi_0(j+1))$ lies in Γ_{j+1}, then the face Γ_{j+1} is invariant. In this case increase j by 1 and go to Step 2. If this condition is not satisfied, go to the next step.

Step 5. Run M iterations of the operator $T_{R,L}$ starting with the vector $\xi_0 = \xi_0(j)$. Compute $\lambda_N = \|T_{R,L}^M \xi_0\|/M$ and

$$\xi_N = \frac{\left(\sum_{n=0}^N \frac{T_{R,L}^n \xi_0}{\lambda_N^n}\right)}{\left\|\sum_{n=0}^N \frac{T_{R,L}^n \xi_0}{\lambda_N^n}\right\|}.$$

Define the face $\Gamma_{j+1} = \{\eta \in \Gamma_j : \eta_i = 0 \text{ if } (\xi_N)_i < \varepsilon\}$. If $\Gamma_{j+1} \neq \Gamma_j$, then go to Step 6, otherwise go to Step 2.

Step 6. (Check that Γ_{j+1} is invariant.) The same as Step 2.

If Γ_{j+1} is invariant, then increase j by 1 and go to Step 2, otherwise go to Stage 2.

Stage 2. Take $\xi_0 = \xi_0(\Gamma_j)$, where Γ_j is the invariant face found at Stage 1. Define $\lambda_N = \|T_{R,L}^M \xi_0\|/M$. Compute

$$\bar{\xi} = \bar{T}_{R,L}^Q \left(\sum_{n=0}^N \frac{T_{R,L}^n}{\lambda_N^n}\right).$$

The vector $\bar{\xi}$ is the desired approximation to the maximal eigenvector.

REMARKS. (1) In this paper we do not consider the important question of convergence in ergodic theorems. This question requires additional analysis.

(2) The question of whether one can determine a satisfying truth assignment from the maximal eigenvector of Maslov's operator is not solved in the general case.

References

1. V. L. Makarov and A. M. Rubinov, *Mathematical theory of economical dynamics and equilibrium*, "Nauka", Moscow, 1973; English transl., Springer-Verlag, Berlin, 1977.
2. I. A. Krass, *Mathematical models of economical dynamics*, "Sovet. Radio", Moscow, 1976. (Russian)
3. K. Yosida, *Functional analysis*, Springer-Verlag, Berlin, 1968.
4. S. Yu. Maslov, Semiotics and Information Science No. 20, VINITI, Moscow, 1983, pp. 3–31. (Russian)
5. M. I. Zakharevich, *Ergodic properties of nonlinear mappings related to models of economical dynamics*, Dokl. Akad. Nauk SSSR **260** (1981), 1298–1301; English transl. in Soviet Math. Dokl. **24** (1981).
6. _____, *Characteristic exponents and the vector ergodic theorem*, Vestnik Leningrad. Univ. **1978**, no. 7 (Ser Mat. Mekh. Astr. Vyp. 2), 28-34; English transl. in Vestnik Leningrad Univ. Math. **11** (1983).
7. _____, *Ergodic properties of Maslov's iterative method*, Third Conf. "Applications of Methods of Mathematical Logic". Abstracts, Tallinn, 1983. (Russian)

Anomalous Properties of Maslov's Iterative Method

I. V. Melichev

ABSTRACT. S. Yu. Maslov's work [6] suggests an iterative method for solving the problem of constructing a satisfying estimate for a 3-cnf and describes a number of experiments run according to it. As noted in [6], the experiments produced good results. Nevertheless, the theoretical analysis of the properties of this iterative method runs into serious problems. First, the statement of the method in [6] involves five parameters which one is invited to adjust for each specific problem, the first three of these parameters being nonnumerical. Second, one has to analyze nonlinear operators of high dimension, which leads to cumbersome calculations.

The experiments described in [6] touch upon only a small section of the spectrum of all possibilities "spanned" by the parameters of the method. We shall consider the version of the method used in these experiments as the basic one. In the present paper we give two negative results concerning this version, viz., examples of formulas on which it is neither correct, nor efficient. After that we give a brief review of possible improvements of the method.

1. Terminology

Following the terminological conventions used in [8], we shall call variables of a propositional logic *positive literals*, and their negations *negative literals*. A pair of literals is said to be *complementary* if it includes both positive and negative literals for the same variable. A *clause* is a formula consisting of (possibly empty) disjunction of literals. A *K-conjunctive normal form* (K-cnf) is a (possibly empty) conjunction of clauses each of which contains at most K literals. An empty cnf is said to be identically true, while a cnf containing an empty clause is said to be identically false.

For a cnf F and a literal a, denote by $F[a]$ the cnf obtained from F by omitting all literals complementary to a, as well as all clauses containing the literal a. The operation of replacing F by $F[a]$ will be called an (elementary) substitution of a in F. A K-cnf is said to be *satisfiable* if there exists a sequence of substitutions resulting in an empty cnf. Then the corresponding collection of literals is called a *satisfying estimate*. It is known that the problem of constructing a satisfying estimate for a 3-cnf is NP-complete.

A substitution of a into F will be called *regular* if F is unsatisfiable, or if F is satisfiable and $F[a]$ is satisfiable. If there is exactly one literal in a complementary pair which gives a correct substitution, then the variable corresponding to it is said to be *rigidly contained* in the formula; otherwise it is said to be *weakly contained*.

We shall number literals by integers. With this in mind, we associate positive integers with the variables contained in the formula, and then assign to each positive literal the number of its variable, and to each negative literal the number of its variable with the "minus" sign. Fix the numbering, and identify literals and their numbers. Accordingly, the propositional negation is sometimes denoted by the "minus" sign.

2. Iterative method

With each literal a contained in a given 3-cnf F we associate a nonnegative real number, called the *deferrence* of a and denoted ξ_a. The vector

$$(1) \qquad (\xi_{-n}, \xi_{-n+1}, \ldots, \xi_{-1}, \xi_1, \xi_2, \ldots, \xi_n)$$

is called an *obstacle* for F (n is the number of variables in the formula F). Choose an initial obstacle ξ^0, and compute

$$(2) \qquad \xi^1 = K_{R,L}\xi^0, \ \xi^2 = K_{R,L}, \ldots, \xi^p = K_{R,L}\xi^{p-1}, \ldots,$$

where $K_{R,L}$ is Maslov's operator (see below). At a certain step, whose number is determined by a special algorithm from the current obstacle, we select a literal a and go from F to $F[a]$ (the literal is selected by a special algorithm from the current obstacle). If $F[a]$ is not empty, we eliminate 1-literal clauses by a sequence of appropriate substitutions. If the resulting cnf is again nonempty and contains no empty clause, we apply the iterative method to it.

The process can have either of two outcomes:

(1) an empty cnf is obtained, and, therefore, a satisfying estimate is constructed;

(2) a cnf with an empty clause is obtained. This may mean that the initial cnf is unsatisfiable; however, it is also possible that the cnf is satisfiable but the iterative algorithm was unable to detect it.

Maslov suggested using in the iterative process (2) an operator of the form

$$(3) \qquad (K_{R,L}\xi)_i = R\xi_i + L \sum_j \min \xi_{F_{j,k}},$$

where $(\cdot)_i$ means the ith component of the vector, $F_{j,k}$ is the literal in the kth position in the clause j of the formula F, j spans the set of all clauses of F containing the literal a while the minimum is taken over all literals of the jth clause except a, and R, L are nonnegative real numbers. In the case of 2-cnf the operation of taking the minimum becomes degenerate and the operator turns into a linear one.

This method involves five parameters: R, L, the criterion for selecting the initial approximation, the algorithm for selecting a step, and the algorithm selecting the literal for substitution. In the experimental runs described in [6], the method has the parameters $R = 0$, $L = 1$, the initial obstacle consists of ones (such an obstacle is said to be *neutral*), the criterion for selecting a step always points to step 1, the algorithm selecting a literal outputs the literal complementary to one of the literals with nonstrictly maximal deferrence. Also, under similar conditions the algorithm was analyzed selecting the literal having the maximum difference of deferrences for complementary pairs (the selected literal is the one for which $\xi_{-a} - \xi_a$ is maximal).

It is not difficult to construct an example in which the method selecting the literal after the first iteration produces an incorrect substitution. Therefore, we shall analyze the behavior of the sequence

(4) $$(K_{R,L}^i \xi^0, \quad i = 0, 1, 2, \ldots).$$

Such an analysis is required if we accept the conjecture stated in [6] that the accuracy of the method increases with the number of iterations.

3. Main properties of the iterative method

It is evident that in the case of 2-cnf the properties of the sequence (4) are defined by the eigenvalue and the position of the eigenvectors of the operator $K_{R,L}$. The same, although to a smaller degree, is true in the case of 3-cnf. However, it is not hard to see that the eigenvectors of Maslov's operator do not depend on the parameters R and L for $L > 0$ (the case $L = 0$ is trivial). The meaning of the coefficients R and L can be easily understood if Maslov's operator is represented in the form

(5) $$K_{R,L} = R \cdot \mathbf{1} + L \cdot K_{0,1}.$$

The operator $K_{0,1}$ will be called the operator of translation. Since we are interested only in the relative values of deferrences, one of the coefficients R, L can be fixed. Set $L = 1$ and let us vary the coefficient R, which we shall call the coefficient of inertia.

We now give (slightly modified) statements of Maslov's theorems.

THEOREM 1. *Let v be a satisfying estimate for formula F and let $\{x\}$ be the set of obstacles such that $x_i = 0$ for $i \in v$. Then $\{x\}$ is an invariant subspace of Maslov's operator for F.*

THEOREM 2. *Let F be a satisfiable 2-cnf. For any $R > 0$ and any initial obstacle with positive coordinates the sequence (4) converges in the direction of an obstacle $\bar{\xi}$ such that $F[\bar{a}]$ is satisfiable, where a is any of the literals having the maximal deferrence in $\bar{\xi}$. (By convergence in the direction is meant the convergence of the sequence $(\xi^i / \|\xi\|, i = 0, 1, \ldots)$.)*

Theorem 2 justifies the correctness of the method in the case of 2-cnf. One would wish to extend it to arbitrary formulas, but, as will be shown below, under natural conditions this is impossible. One would also wish to achieve fast convergence in the direction of the sequence (4). However, as we shall show below, this cannot be guaranteed either.

4. Graph representation of Maslov's operator

In order to provide a more graphical formulation of the following results, we introduce a special representation for operators of Maslov type in the form of weighted graphs. It will be sufficient to consider only the case of 2-cnf. As we have noted, in this case Maslov's operator is a linear one.

Fix a basis in the linear space and consider the matrix $\{M_{ij}\}$ of an arbitrary linear operator in this basis. It is convenient to associate with this matrix an edge-weighted graph, whose vertices correspond to the basis vectors, and each pair of coordinates $\langle i, j \rangle$ corresponds to the edge $i \leftarrow j$ with the weight M_{ij}. Such a representation of linear operators is actively exploited in the theory of electrical circuits [1].

On the other hand, given a 2-cnf F, construct a graph such that its vertices correspond to literals, and for each pair of literals $\langle i, j \rangle$ there is an edge $i \leftarrow j$ whose weight is equal to the number of occurrences of the clause $\bar{i} \vee j$ in formula F (therefore, the edge $i \leftarrow j$ correspond to the implication $i \supset j$). The resulting graph will be called the graph of implication of literals for F.

THEOREM 3. *The graph of translation of Maslov's operator for any 2-cnf coincides with the graph of implications of literals for the same formula.*

The inertia of Maslov's operator can be taken into account by adding to each vertex of the graph a loop with weight equal to the coefficient of inertia.

In practical applications it is convenient to omit edges having weight 0.

5. Example of slow convergence under a special choice of the initial obstacle

Before stating the result on the speed of convergence of the iterative method, let us illustrate the situation by a simplified example. Consider the operator $K_{0,1}$ for the formula whose graph of implications is given in Figure 1 (all edges in this graph have weight 1, which is why the weights of all edges in this figure are omitted). This formula contains $4n + 4$ variables and $4n + 14$ clauses, each consisting of two literals. This formula is true only for $a_{4n+4} = a_{4n+3} = a_{4n+2} = a_{4n+1} = 1$. Let the initial obstacle satisfy the relation

(7)
$$q \leq \xi_k^0 \leq p \quad \text{for } k = -4n-4, -4n-3, \ldots, 4n; \ k \neq 0,$$
$$qu \leq \xi_k^0 \leq pu \quad \text{for } k = 4n+1, 4n+2, 4n+3, 4n+4,$$

where p, q, u are arbitrary coefficients, $0 < q \leq p$, $0 < u$. Then after the ith iteration the deferrences of literals in this formula satisfy the relation

(8) $$qS_{k,i} \leq \xi_k^i \leq pS_{k,i} \quad (k = -4n-4, \ldots, 4n+4; k \neq 0),$$

where

$$S_{k,i} = \begin{cases} 1 & \text{for } -4n \leq k \leq 4n - 4i + 4; \\ u \cdot 3^{[i-n+L\frac{k-5}{4}]} & \text{for } 4n - 4i + 4 < k \text{ and } k < -4n - 1; \\ \dfrac{3^{2i+1} - 1}{2} & \text{for } i \leq 2n, -4n - 4 \leq k \leq -4n - 1; \\ 3^{i-2n-1}\left(\dfrac{3^{2n+2} - 1}{2} + u(i - 2n - 2)\right) & \text{for } i > 2n, -4n - 4 \leq -4 - 1, \end{cases}$$

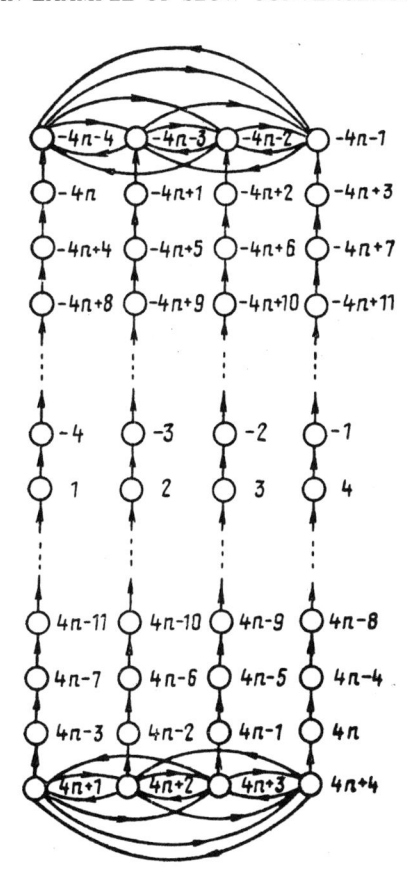

[FIGURE 1 shown here — graph with nodes labeled $-4n-4, -4n-3, \ldots, 4n+4$]

FIGURE 1

(Here $[x]$ is the greatest integer not exceeding x.)

The behavior of the deferrences is given in the table in Section 8.

Evidently, a necessary condition for the correctness of the choice of substitution when selecting a literal by the maximal deferrence is given by the inequality

(9) $$\xi^i_{-4n-4} > \xi^i_{4n+4},$$

which implies $ps_{-4n-4,i} > qS_{4n+4,i}$, or

$$p\left(\frac{3^{n+2}-1}{2} + u(i-2n-1)\right) \cdot 3^{i-2n-1} > q \cdot 3^i \cdot u.$$

This inequality is true only if

(10) $$i - 2n - 1 > \left(\frac{q}{p} - \frac{3}{2u}\right) 3^{2n+1}.$$

The deferrences of other literals are strictly less than the deferrences of the literals with numbers $4n+4, 4n+3, 4n+2, 4n+1$.

Thus, for $u > \frac{3p}{2q}$, if the initial obstacle is selected in the interval (7), then the number of iterations necessary to achieve sufficient accuracy grows exponentially with the size of the formula. The following section contains a generalization of this result.

6. Speed of convergence of the iterative method

THEOREM 4. *For any p, q, $p \geq q > 0$, there is a sequence of 2-cnf $F_1, F_2, \ldots,$ $F_j, \ldots,$ where F_j contains $O\left(\frac{jp}{q}\right)$ clauses, such that for any $R, L > 0$ and for any initial obstacle ξ^0 satisfying the condition $p \leq \xi^0 \leq q$, the iterative method with the operator (3), and the selection tactic*

(A) *by the literal complementary to that having the maximal deferrence, or*

(B) *by the literal a for which $\xi_{\bar{a}} - \xi_a$ is maximal,*

define a regular substitution not earlier than on the third iteration.

In order to prove the theorem in the case $R = 0$ it is sufficient to consider the formula whose graph of implications is given in Figure 2. In this figure every circle stands for the set of four vertices, every arrow means four "parallel" edges, and every double circle denotes a full graph with four vertices. A fragment of this graph is given in Figure 3. The total number of vertices in the graph is $16nu + 8u + 8n + 8$, while the total number of edges is $16nu + 32u + 8u + 28$. If one selects an initial obstacle satisfying the relation $q \leq \xi^0 \leq p$ and takes $u = \left[10\frac{p}{q}\right]$, then for $n > 2\log_3(u+2)$ in the course of the first $\sqrt{3n\frac{q}{p}}$ iterations the vertices with maximum deferrence turn out to be those belonging to the graph S_{II}. All the necessary calculations are done by means of elementary algebra: one considers recurrent relations of the form

$$x_i = 3x_{i-1} + ai^2 \cdot 3^i + bi \cdot 3^i + d.$$

The actual calculations are very cumbersome and are therefore omitted.

In order to prove the theorem in the case $R > 0$, note that

(11) $$K_R^i \xi^0 = (R \cdot \mathbf{1} + K_{0,1})^i \xi^0 = \sum_{l=0}^{i} C_i^l K_{0,1}^l R^{i-l} \xi^0.$$

Since the obstacle $K_{R,1}^i \xi^0$ is a positive linear combination of the obstacles $K_{0,1}^i \xi^0$, the inequalities that are of interest to us can only become stronger.

7. Example of anomalous convergence

Consider the formula

(12) $$\begin{aligned}&(a \vee b \vee \bar{c}) \,\&\, (c \vee d) \,\&\, (\bar{b} \vee d) \,\&\, (\bar{a} \vee d) \,\&\, (x_1 \vee \bar{d}) \\ &\&\, (x_2 \vee \bar{d}) \,\&\, \ldots \,\&\, (x_{625} \vee \bar{d}) \,\&\, (\bar{x}_2 \vee c) \,\&\, \ldots \,\&\, (\bar{x}_{625} \vee c).\end{aligned}$$

In this formula one clause consists of three literals and 1253 clauses consist of two literals. The total number of variables in the formula is 629.

FIGURE 2

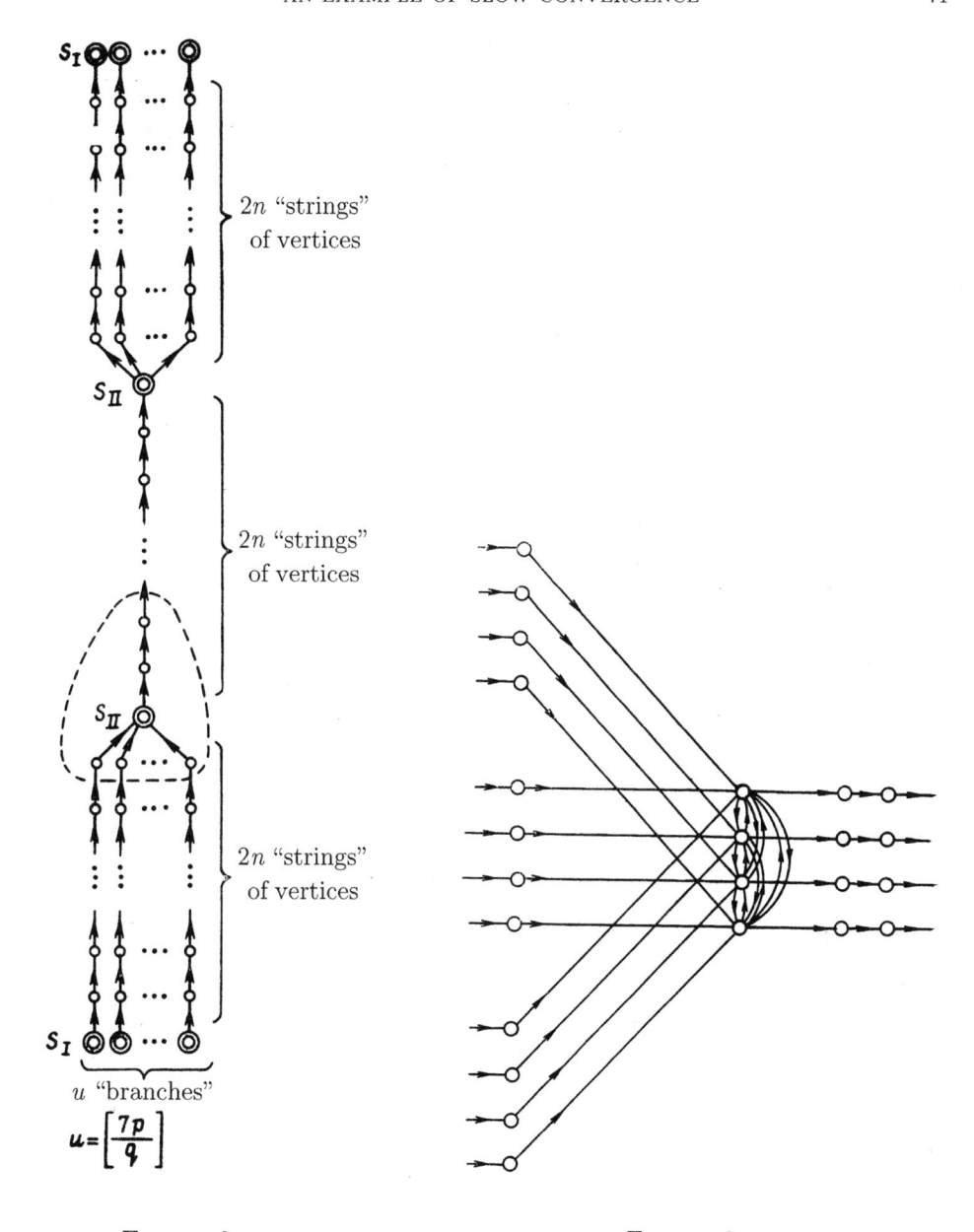

FIGURE 3

Select a neutral initial obstacle. Evidently, the following relations will stay invariant under all iterations:

(13)
$$\xi_{\bar{x}_1} = \xi_{\bar{x}_2} = \ldots = \xi_{\bar{x}_{625}},$$
$$\xi_{x_1} = \xi_{x_2} = \ldots = \xi_{x_{625}},$$
$$\xi_a = \xi_b \leq \xi_{\bar{c}},$$
$$\xi_{\bar{a}} = \xi_{\bar{b}} = \xi_c.$$

Now let us identify literals with equal deferrences and then go to the resulting operator. This operator turns out to be linear, and its matrix is of the form

	d	a,b	$\bar{a}$ $\bar{b}$ c	x	$\bar{d}$	$\bar{x}$	$\bar{c}$
d	R	0	0	625	0	0	0
a,b	1	R	0	0	0	0	0
$\bar{a},\bar{b},c$	0	1	R	0	0	0	0
x	0	0	1	R	0	0	0
$\bar{d}$	0	0	3	0	R	0	0
$\bar{x}$	0	0	0	0	1	R	0
$\bar{c}$	1	0	0	0	0	625	R

It is not hard to compute the principle eigenvalue of this operator $\lambda = R+5$. This eigenvalue corresponds to the eigenvector $(1, \frac{1}{5}, \frac{1}{25}, \frac{3}{625}, \frac{3}{3125}, \frac{4}{25})$. To this eigenvector converges the sequence of iterations (4) for a neutral initial obstacle for $R > 0$ (because the absolute value of the real part of other eigenvector is strictly less). Note that in the limit the literal with maximum deferrence turns out to be d, which defines an incorrect substitution if the selection is made by the maximal deferrence. Indeed, the subformula

(15) $$(a \vee b \vee \bar{c}) \ \& \ (d \vee \bar{a}) \ \& \ (d \vee \bar{b}) \ \& \ (d \vee c)$$

is false for $d = 0$. When the selection is made by the maximal difference of deferrences, the choice falls on the same literal. If we use the iterative process (4) with $R = 0$, then the deferrences define the incorrect substitution at every $(4k+1)$th iteration $(k = 1, 2, \ldots)$.

Thus a direct generalization of Theorem 2 to the case of 3-cnf turns out to be incorrect. Note that every clause of the formula contains at most one negative literal. Therefore, this formula is a Horn's formula, i.e., belongs to a polynomially solvable class [4] (see the table in Section 8, where u is an arbitrary positive number).

8. Discussion

Thus, it turns out that in the general case this version of the iterative method is incorrect and inefficient. The class of formulas it is applicable to efficiently includes neither the class of 2-cnf, nor the class of Horn's formulas. Note that the applicability of the method to Tseitin's formulas [3] is trivial because in Tseitin's formulas all variables are soft ones. Nevertheless, the method is successful in processing many formulas tested in the experiments [6] which do not belong to either of these two classes. So the class of formulas solvable by this method is quite extensive. It would be of interest to obtain a description of this class in terms of the syntactic structure of its formulas.

Another possible direction of further research is to try to modify the method in such a way that it becomes totally correct. As we have already mentioned, Maslov's method involves five parameters which one can vary. We can probably add to them one more parameter — the form of the operator, which, as follows from [5], is also subject to variation. Apparently, the main problem with the operator (3) is that its eigenvalue can be arbitrarily large. All the results discussed above are not true, say, for the operators preserving the simplex

(16) $$\sum_i x_i = \text{const},$$

which can be obtained from Maslov's operator by introducing weight coefficients and negative terms. However, it was not possible to achieve the complete correctness of the method.

TABLE 2

Coefficient	$s_{k,0}$	$s_{k,1}$	$s_{k,2}$	$s_{k,3}$	$\ldots$	$s_{k,2n}$	$s_{k,2n+1}$	$\ldots$	$s_{k,2n+m+1}$ $m \geq 1$
$S_{-4n-4,i} = S_{-4n-3,i} =$ $S_{-4n-2,i} = S_{-4n-1,i} =$	1	4	13	40	$\ldots$	$\frac{3^{2n+1}-1}{2}$	$\frac{3(3^{2n+1}-1)}{2} + u$	$\ldots$	$(\frac{3^{2n+2}-1}{2} + um)3^m$
$S_{-4n,i} = S_{-4n+1,i} =$ $S_{-4n+2,i} = S_{-4n+3,i} =$	1	1	1	1	$\ldots$	u	$3u$	$\ldots$	$3^{m+1}u$
$S_{-4n+4,i} = S_{-4n+5,i} =$ $S_{-4n+6,i} = S_{-4n+7,i} =$	1	1	1	1	$\ldots$	$9u$	$27u$	$\ldots$	$3^{m+3}u$
$\vdots$	$\vdots$	$\vdots$	$\vdots$	$\vdots$		$\vdots$	$\vdots$		$\vdots$
$S_{-4,i} = S_{-3,i} =$ $S_{-2,i} = S_{-1,i} =$	1	1	1	1	$\ldots$	$3^n u$	$3^{n+1} u$	$\ldots$	3^{m+n+1}
$S_{1,i} = S_{2,i} =$ $S_{3,i} = S_{4,i} =$	1	1	1	1	$\ldots$	$3^{n+1} u$	$3^{n+2} u$	$\ldots$	3^{m+n+2}
$\vdots$	$\vdots$	$\vdots$	$\vdots$	$\vdots$		$\vdots$	$\vdots$		$\vdots$
$S_{4n-11,i} = S_{4n-10,i} =$ $S_{4n-9,i} = S_{4n-8,i} =$	1	1	1	u	$\ldots$	$3^{2n-3}u$	$3^{2n-2}u$	$\ldots$	3^{m+2n-2}
$S_{4n-7,i} = S_{4n-6,i} =$ $S_{4n-5,i} = S_{4n-4,i} =$	1	1	u	$3u$	$\ldots$	$3^{2n-2}u$	$3^{2n-1}u$	$\ldots$	3^{m+2n-1}
$S_{4n-3,i} = S_{4n-2,i} =$ $S_{4n-1,i} = S_{4n,i} =$	1	u	$3u$	$9u$	$\ldots$	$3^{2n-1}u$	$3^{2n}u$	$\ldots$	3^{m+2n}
$S_{4n+1,i} = S_{4n+2,i} =$ $S_{4n+3,i} = S_{4n+4,i} =$	u	$3u$	$9u$	$27u$	$\ldots$	$3^{2n}u$	$3^{2n+1}u$	$\ldots$	3^{m+2n+1}

The inefficiency of the method may be related to the conjecture that $P \neq NP$. It is not difficult to boost the efficiency of the method in the case of 2-cnf by using the iterative process

(17) $$\xi^1 = K_{R,L}\xi_0, \ \xi^2 = K_{R,L}^2 \xi_0, \ldots, \xi^p = K_{R,L}^{2^{p-1}} \xi_0, \ldots.$$

However, an attempt to apply this idea in the general case runs into problems, because no efficient way is known to code the powers of nonlinear operators. In any case, the problem of boosting the efficiency of the method is apparently a secondary one with relation to the first two problems.

References

1. V. I. Anisimov, *Topological calculation of electronic circuits*, "Energia", Moscow, 1977. (Russian)
2. M. Garey and D. Johnson, *Computers and intractability: a guide to the theory of NP-completeness*, Freeman, San Francisco, CA, 1979.
3. E. Ya. Dantsin, *Two tautology proof systems based on the splitting method*, Zap. Nauchn. Sem. Leningrad. Otdel. Mat. Inst. Steklov. (LOMI) **105** (1981), 24–44; English transl. in J. Soviet Math. **22** (1983), no. 3.
4. _____, In this volume.
5. V. Ya. Kreinovich, In this volume.
6. S. Yu. Maslov, *Asymmetry of cognitive mechanisms and its implications*, Semiotics and Information Science, No. 20, VINITI, Moscow, 1983, pp. 3–31. (Russian)
7. M. Marcus and H. Minc, *Survey of matrix theory and matrix inequalities*, Allyn and Bacon, Boston, MA, 1964.
8. Ch. Chang and R. Lee, *Symbolic logic and mechanical theorem proving*, Academic Press, New York, 1973.

Amer. Math. Soc. Transl.
(2) Vol. **178**, 1996

Possible Nontraditional Methods of Establishing Unsatisfiability of Propositional Formulas

Yu. V. Matiyasevich

ABSTRACT. A system of nontraditional models is constructed for the NP-complete problem of verifying the satisfiability of propositional formulas.

The methods described in this articles were proposed by the author within the community of experts on mathematical logic in Leningrad in the late 1960s, but were never published for the reason that they were never tested experimentally. The author would like to express his gratitude to the editors for the suggestion of publishing these very raw ideas in this book.

A unifying feature of the methods suggested in this article is that they are inspired by real phenomena in nature. The phenomena are only the starting point for reflection; they give a convenient language, but the author does not propose to realize the corresponding physical processes. We speak only about solving the corresponding differential equations.

We consider the following problem: *find values of Boolean variables $B_1, \ldots, B_l$ for which the formula*

$$(1) \qquad \underset{m=1}{\overset{n}{\&}} (B_{q_{m,1}} \vee B_{q_{m,2}} \vee B_{q_{m,3}})$$

assumes the value "true". Here $1 \leq q_{m,k} \leq 2l$ and B_{l+i} denotes $\neg B_i$.

Method 1: In vitro. Consider a solution containing $2l$ chemicals $M_1, \ldots, M_{2l}$. Reactions among some of these chemicals lead to formation of precipitates, namely, precipitates result from
1. every pair of chemicals M_i and M_{l+i},
2. every triple of chemicals $M_{q_{m,1}}, M_{q_{m,2}}, M_{q_{m,3}}$.

The kinetics of the reactions are described by the concentrations $c_1(t), \ldots, c_{2l}(t)$ of chemicals $M_1, \ldots, M_{2l}$ respectively. These functions satisfy differential equations

$$c_i'(t) = -\alpha c_i^\lambda(t) c_{l+i}^\lambda(t) - \beta \sum c_{q_{m,1}}^\mu(t) c_{q_{m,2}}^\mu(t) c_{q_{m,3}}^\mu(t),$$

©1996 American Mathematical Society

where the sum is taken over all m such that $i = q_{m,k}$ for some k, and $c_{2l+i}(t)$ denotes $c_i(t)$. The parameters α, β, λ, and μ are positive; in real chemical reactions $\lambda = \mu = 1$.

It easy to see that the system tends to a limit equilibrium $c_1(\infty), \ldots, c_{2l}(\infty)$ with $c_i(\infty)c_{l+i}(\infty) = 0$. We shall say that *the solution is nondegenerate* if exactly l chemicals remain. In this case, by setting $B_i := (c_i(\infty) = 0)$, we obtain a truth assignment that satisfies formula (1). When solving the system numerically one can, without waiting for stabilization, set $B_i : (c_i(t) < c_{l+i})$ at each step, and then verify whether this truth assignment satisfies (1). The initial states can be chosen randomly from the unit hypercube.

Degenerate equilibrium states form a set having dimension not greater than $(l-1)$. If formula (1) is satisfiable, then nondegenerate equilibrium states form an l-dimensional subset, i.e., in this case degenerate equilibrium states form a subset of measure zero in it. In order to make certain that this scheme can be applied successfully, an answer to the following question should be found: Under the condition that the formula is satisfiable, does the set of those initial states in the $2l$-hypercube that lead to degenerate equilibria have measure zero? The selection of parameters, especially λ and μ, may turn out to be critical in answering this question; the relation $2\lambda = 3\mu$ should apparently be satisfied.

Method 2: "Creation". In a k-dimensional universe there are p_i atoms of an element a_i, $1 \leq i \leq 2l$. The following gravitational forces act between the atoms:
1. If $|i - j| \neq l$, then an atom of a_i and an atom of a_j attract each other with force $F(r_{ij})$, where r_{ij} is the distance between them.
2. An atom of a_i and an atom of a_{l+i} repel each other with force $G(r_{ij})$.
3. If $i = q_{m,1}$, $j = q_{m,2}$, $k = q_{m,3}$ for some m, then three atoms of a_i, a_j, and a_k respectively repel with forces $H(r_{ij}, r_{jk}, r_{ki})$, $H(r_{jk}, r_{ki}, r_{ij})$, and $H(r_{ki}, r_{ij}, r_{jk})$.

Forces F, G, H decrease to zero at infinity and increase to infinity when atoms come together.

In addition, "the force of the friction on vacuum" may exist; it may depend on velocities and/or higher derivatives.

A "Newton's law" holds in the universe: the acceleration (or a higher derivative) is proportional to the sum of acting forces.

In the evolution of the system, *planets* — gatherings of atoms with only attracting forces — may form. If a planet includes atoms of l elements, then a truth assignment

$$B_i := (\text{the planet has no atoms of } a_i)$$

satisfies formula (1).

The arbitrariness in the choice of "physical laws" which can essentially influence the character and speed of the evolution is even greater than in the preceding method. In order for this scheme to be successful one has to know whether it is possible to construct a universe in such a way that for small quantities p_i some l-element planets will form in a relatively short time provided, of course, that formula (1) is satisfiable.

When the model is interpreted numerically, a planet can be defined as an open ball having either an atom of a_1 or an atom of a_{l+1} as its center, and the radius which is the least number for which atoms of a_j and a_{l+j} for some j fall into its closure.

Method 3: Resonance. There are $2l+1$ pendulums $P_0, \ldots, P_{2l}$ with the same eigenfrequency (the pendulums are considered as one-dimensional oscillators satisfying the differential equation $\ddot{x} = -ax$). The following additional forces depending on velocities act on the pendulums $P_1, \ldots, P_{2l}$:
1. braking forces act on P_i and P_{l+i} when they swing in the same direction, and accelerating forces act on them the rest of the time;
2. braking forces act on $P_{q_{m,1}}, P_{q_{m,2}}, P_{q_{m,3}}$ when they swing in the same direction as P_0, and accelerating forces act on them the rest of the time.

Let $B_1, \ldots, B_l$ be a satisfying assignment for formula (1). Swing the pendulum P_i in phase (respectively, out of phase) with P_0 if B_i is false (respectively, true). In this initial moment, only accelerating forces influence all the pendulums, and one may expect that also in the future most of the time there will be no braking forces, and, out of two pendulums P_i and P_{l+i} one will almost always be in phase with P_0 and the other will almost always be out of phase. Is it true that if formula (1) is satisfiable, then the system will arrive at a similar regime from almost any initial state?

In addition, it is possible to introduce in the system forces of friction and parametric pumping (see, e.g., [**1**]).

References

1. V. I. Arnol'd, *Ordinary differential equations*, "Nauka", Moscow, 1975; English transl., Springer-Verlag, Berlin, 1992.

Dual Algorithms in Discrete Optimization

G. V. Davydov and I. M. Davydova

ABSTRACT. The authors suggest a method of searching for a solution in the discrete optimization problem. Instead of the solution tree used in the branch-and-bound method, they propose to use a dual structure, viz., an object similar to tautology in the propositional calculus.

A dual structure is a set of fragments (not necessarily connected) of all tree branches such that, in order to show that the given solution of the problem is optimal, it is sufficient to verify that the values of the functional cannot be improved on any extension of each fragment from this set to the full branch (of the solution). Generally speaking, the dual structure presents a much more compact device for storing tree fragments than the tree containing the same set of fragments. This fact opens up the possibility of reducing search by directly generating a dual structure.

1. Introduction

As we know, the term "branch-and-bound method" relates to a class of algorithms, and its specific realization is defined by a number of parameters: a way of defining the solution tree, the process of its development, and the way of constructing estimates for partial solutions — subbranches of the tree. The estimates make it possible to reduce search — if the estimate for a subbranch is not better than that for the record one (i.e., the best value of the functional), then there is no need to consider this subbranch any more (nor any of its extensions to the full branch). It will be convenient for us to call such subbranches *closed*. Recall that the problem is solved if all the subbranches of the tree (*subtree* of the complete tree of solutions) are closed. Note that a subbranch is a subset of consecutive arcs in the same branch, the first of which starts from the root of the tree.

One can consider partial solutions of a more general form than a subbranch, namely, not necessarily consecutive subsets of arcs in the same branch. We shall say that a family of partial solutions is a complete family, or a *section* of the tree, if every branch of the complete tree contains a partial solution from the family. In particular, any subtree of a complete tree is a trivial section.

We suggest methods of searching for an optimal solution, under which, instead of subtrees, one constructs sections of arbitrary form. As before, closed partial

©1996 American Mathematical Society

solutions can be defined, and, as before, construction of a closed section, i.e., a section consisting of closed partial solutions, proves that the record is optimal.

One can distinguish between the direct and the dual approaches to solving the discrete optimization problem. The direct approach involves consecutive construction of sections that are more and more closed. The branch-and-bound method realizes this approach for sections having the form of subtrees. Under the dual approach one consecutively constructs closed families of partial solutions [1], which become closer and closer to complete families, i.e., sections.

It has been noted that in specific realizations of the branch-and-bound method it often happens that an optimal solution is quickly found, and most of the computation time is taken by the search required to verify that it is indeed optimal. This is related to the fact that the branch-and-bound method tries to improve the optimal solution by "sounding" quite deeply into the domain of feasible, i.e., non-closed partial solutions. In the methods realizing the dual approach, a proof that a record is optimal is achieved by searching through closed partial solutions which, taken together with partial solutions already found, represent the tree more and more completely. The attempt to improve the optimum is replaced by searching for partial solutions which are "bad", but justify its optimality more and more completely.

The algorithmic realization of the suggested approaches is based on the following duality relation. For each vertex of the tree we define its *branching* as the set of arcs originating at this vertex. Consider the family S of all different branchings in the solution tree. It turns out that any set formed by selecting at least one arc in every branching in S contains some partial solution from a section. This property implies a dual one [1]: the set formed by selecting at least one arc in every partial solution from a section contains some branching from S.

On the basis of the duality relation, one can state direct and dual optimality criteria which make it possible to generate dual algorithms. The first of them generate sections in the solution tree, the second, in the dual tree whose branchings are partial solutions of some section. Combined algorithms are also possible generating sections both in the primal and in the dual tree.

2. Statement of the problem

We shall consider the problem of discrete optimization in the following form.

Given m finite sets B_i, $i \in M = \{1, \ldots, m\}$, and a functional f associating with each vector $\beta \in B_1 \times B_2 \times \ldots \times B_m$ the value $f(\beta)$.

The problem is how to select $\beta^* \in \prod_{i \in M} B_i$ such that

(1) $$f(\beta)^* = \text{opt} = \min\left\{ f(\beta) \mid \beta \in \prod_{i \in M} B_i \right\}.$$

Any vector $\beta = \beta[M] \in \prod_{i=M} B_i$ will be called a *solution*, and, if $f(\beta) = \text{opt}$, an *optimal solution*. A solution $\beta[M]$ defines exactly one element $\beta[i]$ in every set $B_i, i \in M$. Accordingly, we call the sets B_i *alternative sets*, and their elements *alternatives*. Let $I \subseteq M$ be a subset of the index set M. A *partial solution* is a vector $\delta = \delta[I] \in \prod_{i \in I} B_i$ defining alternatives in the sets B_i, $i \in I$. A solution β will be called an *extension* of a partial solution δ if $\beta[I] = \delta[I]$. We shall assume, for the sake of simplicity, that the sets B_i, $i \in M$, are pairwise distinct. Then the notation $\beta \supseteq \delta$ has the usual set-theoretical meaning for any partial solution δ and an extension β.

Many problems of discrete programming can be presented in the form (1).

For example, in the multidimensional knapsack problem [2] every set B_i, $i \in M$, consists of two alternatives: i — object i is selected, $\bar{i}$ — object i is not selected. If a solution β turns out to be admissible, i.e., if it satisfies the conditions of the knapsack problem, then the value of the functional $f(\beta)$ is equal to the value of the objective function on this solution. If the solution β is inadmissible, the value $f(\beta)$ is set equal to a sufficiently large number (or a sufficiently small one, if in (1) we consider max instead of min).

The multivariant location problem can also be reduced to the form (1). The role of the sets $B_i, i \in M$, is played by the sets of possible construction schemes in each of the locations $i \in M$. The value of the functional $f(\beta)$ defines the expenses involved according to the construction plan, i.e., the solution β.

In a similar way, one can reduce the travelling salesman problem, as well as other NP-complete problems, to the form (1).

Suppose that for each partial solution δ an *estimate* $\text{est}(\delta)$ is defined satisfying the condition

(2) $$\text{est}(\delta) \leq f(\beta) \quad \text{for any } \beta \supseteq \delta.$$

Suppose that a solution β and a value of the functional $\text{rec} = f(\beta)$ are known. Partial solutions δ satisfying the condition

(3) $$\text{est}(\delta) \geq \text{rec}$$

will be called *closed*. By virtue of (2) the value of rec cannot be made smaller on any extension β of the closed partial solution δ. This is why in the branch-and-bound method all extensions of closed partial solutions are discarded. Finally, all partial solutions that were considered form a trivial section of the tree whose branchings are alternative sets.

In order to describe nontrivial sections consisting of closed partial solutions and the properties of such sections, let us introduce the following notions.

3. Dual structures

Let A be a finite set of elements (alphabet), and B and D fixed families of nonempty subsets of A. We say that B is *dual* to D if any set α having nonempty intersection with every set ω of the family B contains a set $\delta \in D$.

EXAMPLE. Let $A = \{a, b, c, d, e, f, g, h\}$. We shall denote subsets forming families by columns, and write all the columns of family B to the left of a vertical line, and all the columns of family D to the right of the line. Consider a pair of families

(4) $$(B \mid D) = \begin{pmatrix} & & e & & c & & & & & \\ a & \boldsymbol{b} & f & d & d & a & a & c & \boldsymbol{b} & g \\ \boldsymbol{h} & c & g & \boldsymbol{b} & e & b & g & f & \boldsymbol{h} & h \end{pmatrix}.$$

Consider, for example, the set $\alpha = \{h, b, e\}$. In the columns of family B the elements of the nonempty intersection of each column in B with the set α are shown in bold. The column $\delta = \{b, h\}$ shown in bold in D is contained in the set α.

DUALITY THEOREM. *If family B is dual to D, then D is dual to B.*

PROOF. Suppose that the statement of the theorems is false. Then there exists a set γ having a nonempty intersection with every set $\delta \in D$ and containing none of the sets $\omega \in B$. Consider the complementary set $\alpha = A \setminus \gamma$. In each set $\omega \in B$ there are elements from α because γ includes none of the sets $\omega \in B$. Thus, α has a nonempty intersection with every $\omega \in B$, but contains none of the sets $\delta \in D$ because every $\delta \in D$ has elements from $\gamma = A \setminus \alpha$ that do not belong to α. This contradicts the duality of B and D. The theorem is proved.

A pair of families $(B \mid D)$ will be called a *structure*, and, if B is dual to D, then $(B \mid D)$ will be called a *dual structure*. The sets belonging to B and D will be called *columns*. A *path* is a set π formed by selecting one element in every column of the family.

It is easy to see that the structure (4) is a dual one. The set $\alpha = \{h, b, c\}$ is an example of a path in the family B, and the set $\gamma = \{c, a, g, h\}$ is an example of a path in the family D. Note that the set γ is a path if we select either h or g in the last column $\{g, h\}$ of D.

Since each path has a nonempty intersection with every column of the family, then in any dual structure $(B \mid D)$ any path in B (respectively, D) contains a column in D (respectively, B).

In order to clarify one of possible interpretations of the dual structure notion, let us construct a disjunctive normal form from (4) by replacing each literal in D with its negation, and considering every column as a conjunction and the families as disjunctions

(5)
$$(a \ \& \ h) \vee (b \ \& \ c) \vee (e \ \& \ f \ \& \ g) \vee (d \ \& \ b)$$
$$\vee (\neg c \ \& \ \neg d \ \& \ \neg e) \vee (\neg a \ \& \ \neg b) \vee (\neg a \ \& \ \neg g)$$
$$\vee (\neg c \ \& \ \neg f) \vee (\neg b \ \& \ \neg h) \vee (\neg g \ \& \ \neg h).$$

It easy to check that *structure (4) is dual if and only if the disjunctive normal form (5) is a tautology.*

Dual structures were introduced in [4] under the name of closed diagrams. If, in addition to duality, every column $\omega \in B$ has a nonempty intersection with every column $\delta \in D$, then the structure $(B \mid D)$ forms a dual system of sets in the sense of [5].

An equivalent statement of duality is given by the following lemma [6].

LEMMA 1. *A structure $(B \mid D)$ is dual if and only if any path π in B intersects any path ρ in D.*

NECESSITY. Let $(B \mid D)$ be a dual structure. Then any path π in B contains a column $\delta \in D$. By definition, any path ρ in D has a nonempty intersection with any column in D, and so $\rho \cap \delta \neq \emptyset$. Since $\pi \supseteq \delta$, we have $\rho \cap \pi \neq \emptyset$.

SUFFICIENCY. Suppose that $\rho \cap \pi \neq \emptyset$ for any paths π in B and ρ in D, and let us prove that B is dual to D. Let γ be an arbitrary finite set, and $\gamma \cap \beta \neq \emptyset$ for any column $\beta \in B$. Choose an element in each intersection $\gamma \cap \beta$ (and, therefore, in each column β) and form a path π consisting of the elements selected. Let us show that there exists a column $\delta \in D$ such that $\pi \supseteq \delta$. Indeed, otherwise each column in D contains an element that does not belong to π. Then we can form a path ρ in D such that $\pi \cap \rho = \emptyset$, which contradicts the assumption.

Let us now clarify the meaning of the dual structure in problem (1).

4. Direct and dual optimization criteria

Consider as our family B the family $B = \{B_i, i \in M\}$ of all alternative sets, and as D the family of partial solutions.

The family D of partial solutions is said to be *complete* if the structure $(B \mid D)$ is dual.

Consider two types of trees: a *direct tree* whose branchings are defined by columns in B, and a *dual tree* whose branchings are defined by columns in D.

EXAMPLE. Let the dual structure $(B \mid D)$ be of the form

$$\begin{pmatrix} a & b & e & | & a & a & d & d \\ d & c & f & | & b & c & e & f \end{pmatrix}. \tag{6}$$

Then examples of a direct and a dual tree are given by:

(A) DIRECT TREE   (B) DUAL TREE

These trees are subtrees of the corresponding complete trees:

(A) COMPLETE DIRECT TREE   (B) COMPLETE DUAL TREE

The duality of B to D implies that any path in P contains a column in D. But any path in B is a branch in a complete direct tree. Therefore the complete direct family D of partial solutions is a section of the complete direct tree mentioned in the introduction. Similarly, the duality of D to B means that family B is a section of the complete dual tree.

A family D of partial solutions is said to be *closed by the record* rec if

(a) each partial solution $\delta \in D$ is closed, i.e., $\text{est}(\delta) \geq \text{rec}$, and

(b) the family D contains a partial solution δ_r such that there exists an extension $\beta_r \supseteq \delta_r$ with $f(\beta_r) = \text{rec}$.

The definition implies, in particular, that $\text{est}(\delta_r) = \text{rec}$. Indeed, because of (a), we have $\text{est}(\delta_r) \geq \text{rec}$, and property (2), which defines the estimate, implies that $\text{est}(\delta_r) \leq f(\beta_r) = \text{rec}$. Comparing the two inequalities, we obtain that $\text{est}(\delta_r) = \text{rec}$.

A family of partial solutions is said to be *closed* if it is closed by some record. Note that if D_1 is closed by the record rec_1 and D_2 is closed by the record rec_2, then $D = D_1 \cup D_2$ is closed by the record $\text{rec} = \min\{\text{rec}_1, \text{rec}_2\}$. In other words, the union of closed families is again a closed family. Adding a closed partial solution to a closed family does not change its property of being closed.

In order to solve problem (1) it is sufficient to be able to construct complete closed families. This is justified by the following assertion.

DIRECT OPTIMALITY CRITERION. *If a complete family D is closed by the record $\text{rec} = f(\beta^*)$, then $\text{rec} = \text{opt}$ and β^* is an optimal solution.*

Indeed, the completeness of D means the duality of the structure $(B \mid D)$. Since any solution β forms a path in B, then the duality of B and D implies that there exists $\delta \in D$ such that $\beta \supseteq \delta$. The fact that δ is closed and the definition of estimate (2) imply that $f(\beta) \geq \text{est}(\delta) \geq \text{rec}$. This means that $\text{opt} \geq \text{rec}$. Property (b) in the definition of a closed family guarantees the existence of a solution β^* on which rec is realized, i.e., $\text{opt} = \text{rec} = f(\beta^*)$ and β^* is an optimal solution.

The direct optimality criterion is based on the duality of B to D. What does the duality of D to B mean in terms of our problem? In order to answer this question, consider a family D which is closed by the record rec under the assumption $\text{rec} > \text{opt}$. Let β^* be an arbitrary optimal solution. Then for any extension β of a partial solution $\delta \in D$ the following inequalities hold:

$$f(\beta) \geq \text{est}(\delta) \geq \text{rec} > \text{opt} = f(\beta^*).$$

Therefore, no $\delta \in D$ can be extended to β^*, i.e., $\beta^* \not\supseteq \delta$. Since the partial solution is $\delta = \delta[I]$, the fact that $\beta^*[M] \not\supseteq \delta[I]$ implies that there is an index $i = i(\delta, \beta^*)$ in I that is represented by different alternatives, i.e., $\delta[i(\delta, \beta^*)] \neq \beta^*[i(\delta, \beta^*)]$. Thus, in every partial solution $\delta \in D$ one can select a "bad" alternative $\delta[i(\delta, \beta^*)]$ which does not belong to the optimal solution β^*. The path $\pi(\beta^*) = \{\delta[i(\delta, \beta^*)] \mid \delta \in D\}$ of "bad" alternatives in D will be called a *refutation with respect to β^**. If the value under consideration is $\text{rec} = f(\beta_r)$, then the solution β^* is a counterexample which is a refutation with respect to some optimal solution. By a *refutation* we mean a path which is a refutation with respect to some optimal solution.

Thus, we have shown that if $\text{rec} > \text{opt}$, then there exists a refutation in D. This makes it possible to state another criterion of optimality.

DUAL OPTIMALITY CRITERION. *If none of the paths in a family D which is closed by the record rec is a refutation, then $\text{rec} = \text{opt}$.*

Indeed, if $\text{rec} > \text{opt}$, then, as was demonstrated, there is a refutation in D.

The practical application of the dual criterion is facilitated by the following property.

NECESSARY REFUTATION CONDITION. *Suppose that the path π is a refutation in a closed family D. Then π contains none of the alternative sets $B_i, i \in M$.*

Indeed, the path $\pi = \pi(\beta^*)$ is a refutation with respect to β^*. This means that the path π includes no alternatives forming the solution β^*, i.e., $\beta^*[i] \notin \pi$ for each $i \in M$. Since $\beta^*[i] \in B_i, i \in M$, then π contains none of the sets $B_i, i \in M$.

A sufficient condition for the dual optimality criterion to be satisfied is the completeness of the closed family D.

Indeed, any path in a complete family D contains an alternative set $B_i, i \in M$, and hence, in view the necessary condition, it is not a refutation.

Let us illustrate the dual optimality criterion by the example of 2-element alternative sets.

Suppose that a family D closed by the record $\mathrm{rec} = f(\beta_r)$ is constructed. In order to verify the optimality of the solution β_r, let us try to find a counterexample rejecting its optimality. Let π be a path in D containing no alternative set. If no such path exists, then β_r is an optimal solution. Otherwise π is a candidate for refutation. Construct a partial solution $\delta[I]$ that could provide a counterexample as follows: for each alternative the element $a \in \pi$ defines an index $i \in M$ such that $\{a, \delta[i]\}$ is an alternative set, the set I is formed by the indices thus selected. If it turns out that $\mathrm{est}(\delta) \geq \mathrm{rec}$, then the extensions δ are not counterexamples, and the number of possible counterexamples decreases. Adding δ to D, we obtain a closed family that is closer to a complete one. If $\mathrm{est}(\delta) < \mathrm{rec}$, the partial solution δ may be extended to obtain a counterexample. The way to do it will be described below.

EXAMPLE. Let $m = 5$, $B_i = \{i, \bar{i}\}$, $i = 1, \ldots, 5$, i.e., the family B is of the form

$$B = \begin{pmatrix} 1 & 2 & 3 & 4 & 5 \\ \bar{1} & \bar{2} & \bar{3} & \bar{4} & \bar{5} \end{pmatrix}.$$

Suppose that the solution $\beta_r = \{\bar{1}, \bar{2}, \bar{3}, 4, 5\}$ is known, and that the value of the functional is $\mathrm{rec} = f(\beta_r)$. Suppose that it is also known that $\mathrm{est}(\delta_r) = \mathrm{rec}$ for the partial solution $\delta_r = \{\bar{1}, \bar{2}, \bar{3}\}$, and that the partial solutions $\delta_1 = \{\bar{1}, 2\}, \delta_2 = \{\bar{2}, 3\}, \delta_3 = \{1, \bar{3}\}$ satisfy the condition $\mathrm{est}(\delta_k) \geq \mathrm{rec}, k = 1, 2, 3$. Then the partial solutions $\delta_r, \delta_1, \delta_2, \delta_3$ form the closed family

$$D = \begin{pmatrix} \bar{1} & & & 1 \\ \bar{2} & \bar{1} & 2 & \\ \bar{3} & 2 & \bar{3} & \bar{3} \end{pmatrix}.$$

Consider the path $\pi = \{\bar{1}, \bar{2}, \bar{3}\}$ in D. It contains no alternative sets and is a candidate for refutation. By adding the new partial solution $\delta_4 = \{1, 2, 3\}$ (a possible counterexample), we obtain that any extension of the path π through the column δ_4 contains an alternative set. If $\mathrm{est}(\delta_4) \geq \mathrm{rec}$ and no other paths are candidates for refutation, then the problem is solved. In our case it is easy to check that the structure

(7)
$$\begin{pmatrix} 1 & 2 & 3 & \bar{1} & & & 1 & \\ & & & \bar{2} & \bar{1} & 2 & & 2 \\ \bar{1} & \bar{2} & \bar{3} & \bar{3} & 2 & \bar{3} & \bar{3} & 3 \end{pmatrix}$$

is a dual one. Adding two more columns $\{4, \bar{4}\}$ and $\{5, \bar{5}\}$ to the left-hand family does not destroy the duality. Thus no other paths are candidates for refutation, and for $\mathrm{est}(\delta_4) \geq \mathrm{rec}$ the problem would be solved.

Let us now describe informally how one can construct complete closed families, i.e., dual structures that have a closed family on the right-hand side and alternative sets on the left-hand side.

First we consider a structure that is known to be dual, which we shall call a *template*. By substituting the alternatives of our problem for the letters of the alphabet of the template, we obtain a dual structure $(X \mid Y)$, called an *interpretation* of the template. The columns in the structure $(X \mid Y)$ consist of alternatives. A column $x \in X$ (respectively, $y \in Y$) is said to be *closed* if it contains an alternative set (respectively, a closed partial solution). Note that the closeness of columns in Y is defined only if the functional f, the estimate est, and the record rec are known. A path in X (respectively, Y) is said to be *closed* if it contains a closed column in Y (respectively, X). If all columns in the interpretation $(X \mid Y)$ are closed, then the optimality criteria imply that the problem is solved. If not, then one can select a structure $(X_1 \mid Y_1)$, called a *complement*, consisting of all nonclosed columns $X_1 \subset X$ and $Y_1 \subset Y$.

If $X_1 = \emptyset$, then in order to find an optimal solution it is sufficient to examine and estimate extensions of partial solutions belonging only to the set Y_1.

By considering different interpretations of templates, one can obtain different complements which can then be used to construct dual structures with closed columns.

5. Templates

For a template one can take any dual structure. The duality of some of the structures can be verified by checking all their paths on either side of the prospective template. New dual structures can also be generated from those already constructed using certain rules. This will be described in detail in the appendix to the article.

For example, it easy to demonstrate by direct examination of all paths, either in X or in Y, that the following structures are dual:

(a) $\qquad (X \mid Y) = (a \mid a),$

(b) $\qquad (X \mid Y) = \begin{pmatrix} a_1 & a_2 & \cdots & a_n & \begin{vmatrix} a_1 \\ a_2 \\ \vdots \\ a_n \end{vmatrix} \end{pmatrix},$

(c) $\qquad (X \mid Y) = \begin{pmatrix} a_{11} & a_{21} & & & & & \\ a_{12} & a_{22} & \begin{vmatrix} a_{11} \\ \\ a_{21} \end{vmatrix} & a_{12} & \cdots & a_{1n_1} & a_{22} & \cdots & a_{2n_2} \\ \vdots & \vdots & & & & & \\ a_{1n_1} & a_{2n_2} & & & & & \end{pmatrix},$

(d) $\qquad (X \mid Y) = \begin{pmatrix} & a_1 & & & & & \\ & a_2 & \begin{vmatrix} a & a & & a \\ a_1 & a_2 & \cdots & a_n \end{vmatrix} \\ a & \vdots & & & & \\ & a_n & & & & & \end{pmatrix},$

(e) $\quad (X \mid Y) = \begin{pmatrix} a & a & b & b & c & c & d & a \\ d & e & e & f & f & g & g & c & e & g \end{pmatrix}$,

(f) $\quad (X \mid Y) = \begin{pmatrix} a & c & e & a & g & & & \\ a & c & e & c & b & a & c & e \\ b & d & g & e & d & g & b & d \end{pmatrix}$.

Let $(X \mid Y)$ be a template in the alphabet Z, and A an alphabet. An *interpretation* of the template $(X \mid Y)$ in the alphabet A is a mapping $\varphi: Z \to A$ associating with each element $z \in Z$ an element $\varphi(z) \in A$. The mapping φ maps each column x of the template $(X \mid Y)$ into the column $\varphi(x)$. As usual, $\varphi(X) = \{\varphi(x) \mid x \in X\}$ is the image of the family X, while $\varphi(Y) = \{\varphi(y) \mid y \in Y\}$ is the image of the family Y. The pair $(\varphi(X) \mid \varphi(Y))$ will also be called an interpretation of the template.

LEMMA 2. *$(\varphi(X) \mid \varphi(Y))$ is a dual structure.*

PROOF. Any path π in $\varphi(X)$ is the image of some path π' in X, i.e., $\pi = \varphi\pi'$. The same holds for any path ρ in $\varphi(Y)$, i.e., $\rho = \varphi(\rho')$, where ρ' is a path in Y. Since $(X \mid Y)$ is a template, then, by Lemma 1, $\pi' \cap \rho' \neq \emptyset$. We have $\pi \cap \rho = \varphi(\pi' \cap \rho') \neq \emptyset$, which, by the same Lemma 1, means that $(\varphi(X) \mid \varphi(Y))$ is a dual structure.

EXAMPLE OF AN INTERPRETATION. Consider the template (f) defined in the alphabet $Z = \{a, b, c, d, e, g\}$. Set $A = \{1, \bar{1}, 2, \bar{2}, 3, \bar{3}, 4, \bar{4}, 5, \bar{5}\}$ and consider the interpretation φ defined by $\varphi(a) = 1, \varphi(b) = \bar{1}, \varphi(c) = 2, \varphi(d) = \bar{2}, \varphi(e) = 3, \varphi(g) = \bar{3}$. Then the interpretation of the template (f) is the already known dual structure (7):

$$\begin{pmatrix} 1 & 2 & 3 & 1 & \bar{1} & & & \\ 1 & 2 & 3 & 2 & \bar{2} & \bar{1} & 2 & 1 \\ \bar{1} & \bar{2} & \bar{3} & 3 & \bar{3} & 2 & 3 & \bar{3} \end{pmatrix}.$$

Let $(B \mid D)$ be the tested structure in the alphabet A. We now state a sufficient condition for its duality.

SUFFICIENT DUALITY CONDITION. *Let $(B \mid D)$ be a structure, $(X \mid Y)$ a template, and φ an interpretation of $(X \mid Y)$ such that each column in $\varphi(Y)$ contains a column in D. Then in order for the structure $(B \mid D)$ to be dual it is sufficient that each column in $\varphi(X)$ contains a column in B.*

Indeed, any path π in D contains a path ρ in $\varphi(Y)$. By assumption, there is relation between $\varphi(Y)$ and D that can be graphically represented as follows:

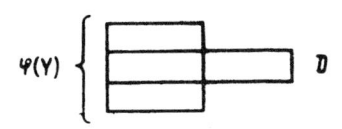

But since $(\varphi(X) \mid \varphi(Y))$ is a dual structure, the path ρ contains a column x in $\varphi(X)$. If the sufficient condition is satisfied, then x contains a column ω in B. Thus we have $\pi \supseteq \rho \supseteq x \supseteq \omega$, whence it follows that $\pi \supseteq \omega$, which means that the structure $(B \mid D)$ is dual.

Let us check the duality of the structure

$$(8) \qquad (B \mid D) = \begin{pmatrix} 1 & 2 & 3 & 4 & 5 & 1 & \bar{1} \\ \bar{1} & \bar{2} & \bar{3} & \bar{4} & \bar{5} & 2 & \bar{2} \\ & & & & & & \bar{3} & 2 & 3 & 1 \end{pmatrix}.$$

Take the template (f) with the interpretation as in the preceding example. We obtain the structure $(\varphi(X) \mid \varphi(Y))$ of the form (7) satisfying the sufficient condition: each column in $\varphi(Y)$ contains a column in D and each column in $\varphi(X)$ coincides with one of the columns in B. Thus the structure (8) is a dual one.

The interpretation φ discussed in the sufficient duality condition breaks all paths in D into classes. One class includes all the paths containing a fixed column in $\varphi(X)$. This makes it possible to reduce the problem of verifying that all paths in a given class are closed to the verification of a single column in $\varphi(X)$ representing this class. If all the columns in $\varphi(X)$ turn out to be closed, so are all paths in D.

For the sake of simplicity we shall consider only a partial case of problem (1): all alternative sets consist of two elements, and $B_i = \{i, \bar{i}\}, i \in M$.

In the appendix to this article the general case is reduced to this partial problem.

In the case of two-element alternative sets, if the sufficient duality condition is not satisfied, there is a very simple way to extend the tested structure to a dual one.

Suppose that the following condition holds.

CONDITION (∗). $(B \mid D)$ is the tested structure, $B = \{B_i \mid i \in M\}$, and in the interpretation of the template $(X \mid Y)$ every column in $\varphi(X)$ contains a column in D, but there exists at least one column in $\varphi(X)$ containing no column from B.

Denote by U the family of all such columns in $\varphi(X)$.

In an arbitrary column u, let us replace each alternative with its complement, i.e., the alternative i by $\bar{i}$, and $\bar{i}$ by i. The resulting column will be called the *inversion of the column* u and denoted by $I(u)$. The *inversion of the family* U is the family $I(U) = \{I(u) \mid u \in U\}$.

LEMMA 3. *If Condition* (∗) *is satisfied, then the structure* $(B \mid DI(U))$ *is dual.*

PROOF. Consider an arbitrary path π in the family $DI(U)$. Let $\pi_D = \pi \cap D$. It follows from Condition (∗) and the duality of the interpretation $(\varphi(X) \mid \varphi(Y))$ that there exists a column $x \in \varphi(X)$ such that $\pi_D \supseteq x$. If $x \notin U$, then x contains a column $\omega \in B$, i.e., $\pi \supseteq \pi_D \supset x \supset \omega$, and π is a closed path in $DI(U)$.

Let $x \in U$. Consider the subpath $\pi' = \pi \cap (D \cup I(x))$. The subpath π' is the completion of the path π_D by an element a from the column $I(x)$. Since $\pi_D \supseteq x$, then π_D must contain the alternative $I(a)$ which together with a forms the alternative set $\{a, I(a)\}$. Thus $\pi \supseteq \pi' \supseteq \{a, I(a)\}$, and π is a closed path in $DI(U)$.

We have shown that any path π in $DI(U)$ is closed both for $x \notin U$ and $x \in U$, i.e., the structure $(B \mid DI(U))$ is a dual one. The lemma is proved.

EXAMPLE OF APPLICATION OF LEMMA 3. Consider the structure of the form

(9)
$$(B \mid D) = \begin{pmatrix} 1 & 2 & 3 & 4 & 5 & \begin{vmatrix} 1 \\ 3 \end{vmatrix} & \bar{1} & \bar{3} \\ \bar{1} & \bar{2} & \bar{3} & \bar{4} & \bar{5} & \bar{5} & \bar{3} & \bar{5} \end{pmatrix}.$$

Consider the template (e)

$$\begin{pmatrix} a & a & b & b & c & c & d & \begin{vmatrix} a \\ b & d & f \\ d & e & e & f & f & g & g & c & e & g \end{vmatrix} \end{pmatrix}$$

in the alphabet $Z = \{a, b, c, d, e, f, g\}$. Define the interpretation φ by the relations $\varphi(a) = 1, \varphi(b) = 3, \varphi(c) = 5, \varphi(d) = \bar{1}, \varphi(e) = \bar{3}, \varphi(f) = \bar{3}, \varphi(g) = \bar{5}$. The resulting interpretation of the template is of the form

$$\begin{pmatrix} 1 & 1 & 3 & 3 & 5 & 5 & \bar{1} & \begin{vmatrix} 1 \\ 3 \end{vmatrix} & \bar{1} & \bar{3} \\ \bar{1} & \bar{3} & \bar{3} & \bar{3} & \bar{3} & \bar{5} & \bar{5} & \bar{5} & \bar{3} & \bar{5} \end{pmatrix}$$

Condition (∗) is satisfied for the family

$$U = \begin{pmatrix} 1 & 5 & \bar{1} \\ 3 & 3 & 5 \end{pmatrix}.$$

By the definition of inversion,

$$I(U) = \begin{pmatrix} \bar{1} & \bar{5} & 1 \\ 3 & 3 & 5 \end{pmatrix}.$$

Lemma 3 states that the structure

$$\begin{pmatrix} 1 & 2 & 3 & 4 & 5 & \begin{vmatrix} 1 \\ 3 \end{vmatrix} & \bar{1} & \bar{3} & \bar{1} & \bar{5} & 1 \\ \bar{1} & \bar{2} & \bar{3} & \bar{4} & \bar{5} & \bar{5} & \bar{3} & \bar{5} & 3 & 3 & 5 \end{pmatrix}$$

is a dual one.

An application of Lemma 3 to problem (1) makes it possible to complete the closed family D to the dual structure $(B \mid DD')$ by adding the family D' of new partial solutions. If every partial solution $\delta \in D'$ turns out to be closed (est$(\delta) \geq$ rec) in the sense of either the old value of rec, or a new one obtained on some extension of $\delta \in D'$, then the problem is solved. Let $K \subset D'$ be the family of all *nonclosed* partial solutions δ, i.e., such that est$(\delta) <$ rec. The family K is a part of the section (DD') of the direct tree. If all extensions of partial solutions $\delta \in K$ are examined, the problem is solved (each branch of the direct tree either contains a closed partial solution, or is an extension of a nonclosed partial solution $\delta \in K$). The faster one can construct a section with the empty family K, the faster the problem is solved.

In what follows, families K will be called complements. The term tries to emphasize that in order to complete the examination of the whole tree, it is sufficient to consider branches that are extension of only those nonclosed partial solutions that form K.

6. Complements

As before, $B = \{B_i \mid i \in M\}$. In addition to the notion of a nonclosed partial solution, let us describe a *nonclosed set*, i.e., a set containing no alternative set.

A family R of nonclosed partial solutions (respectively, the set L of nonclosed sets) is said to be a *direct* (respectively, *dual*) *complement*, if there exist a nonempty closed family D and a nonempty family $G \subseteq B$ of alternative sets such that the structure $(G \mid RD)$ (respectively, $(G \mid RD)$) is a dual one.

We shall require the following simple property of dual structures.

LEMMA 4. *If $(X \mid Y)$ is a dual structure, then so is the structure $(X'X \mid YY')$ for any families of columns X' and Y'.*

Indeed, any path in $X'X$ contains a path in X, while in view of the duality of $(X \mid Y)$ it also contains a column in Y, i.e., from YY'

If both the direct and the dual complements are empty, then the closed family D is complete. Indeed, in this case $(B \mid D)$ is a dual structure and $G \subseteq B$. Therefore, by Lemma 4, the structure $(B \mid D)$ is also a dual one, and so D is a complete family.

Now consider examples of trivial direct and dual complements. Suppose that $\alpha = \{\alpha_1, \alpha_2, \ldots, \alpha_k\}$ is a finite set. The family of sets $\{\{\alpha_1\}, \{\alpha_2\}, \ldots, \{\alpha_k\}\}$ will be called the *transpose* to α and denoted α^T (the column α is considered as a row of one-element columns).

The family $B_i^T = (i, \bar{\imath})$ is a *trivial direct complement* for any $i \in M$ if $\{i\}$ and $\{\bar{\imath}\}$ are nonclosed partial solutions for some value of the record rec. Indeed, it is easy to see that the structure $\left(\dfrac{i}{\bar{\imath}} \;\middle|\; i\bar{\imath}\right)$ is dual. By Lemma 4, adding on the right a family D closed by the record rec does not destroy the duality and results in a structure of the form $(G \mid RD)$, where $G = B_i$, $R = B_i^T$.

In order to give the definition of a dual complement, consider a partial solution $\delta = \{\delta_1, \delta_2, \ldots, \delta_k\}$ that is closed for some record rec. Then δ^T is a *trivial dual complement*. Indeed, first, every column δ_j, $j = 1, \ldots, k$, is nonclosed because, as it is a one-element set, it cannot contain an alternative set. Second, it is easy to check that the structure $(\delta^T \mid \delta)$ is dual. Third, by Lemma 4, adding to this structure a family B on the left, and on the right a family D' closed by the record rec does not destroy the duality and results in a dual structure of the form $(LG \mid D)$, where $L = \delta^T$, $G = B$, and $D = D' \cup \{\delta\}$ is a closed family.

We have already explained the meaning of nonemptiness for a direct complement. A dual complement L is the nonclosed part of a section in the dual tree whose branchings are defined by partial solutions from D. In order to verify that every branch of this tree is closed, it is sufficient to examine extensions to full branches of only the nonclosed sets in the complement L.

In order to obtain either a direct or a dual complement, it is convenient to generate complements containing a small number of long columns. This is facilitated by the following properties of dual structures.

We shall say that a family S_1 *absorbs the family* S_2 if each column in S_1 contains a column in S_2.

We shall say that a column α *absorbs the column* β if the family α^T absorbs the family β^T, i.e., if $\alpha \subseteq \beta$.

LEMMA 5 (ABSORPTION RULE). *Suppose that $(X \mid Y)$ is a structure, α, β are columns in Y, and α absorbs β. Then the structure $(X \mid Y)$ is dual if and only if $(X \mid Y \setminus \{\beta\})$ is dual.*

PROOF. Let the structure $(B \mid D)$ be dual. Since any path in X contains a path in Y, one can replace the column β with the column α without destroying the duality: a path containing β also contains α. Of two equal columns, viz., the column α and the column β reduced to α, one can be dropped without affecting the duality of X to Y.

If the structure $(X \mid Y)$ is not dual, then dropping any column from Y can only make paths in Y shorter, which does not turn the structure into a dual one. The lemma is proved.

Lemma 5 enables one to omit absorbed columns from closed families and complements. We can also eliminate complements for which there exist others that absorb them. Indeed, suppose that a complement K_1 absorbs a complement K_2. In order to examine all the solutions, it is sufficient to consider extensions of the columns from some complement. But the set of extensions of K_1 is contained in the set of extensions of K_2. The complement K_1 is, therefore, preferable: in general, fewer extensions have to be examined for it than for K_2.

The following lemmas and remarks are instrumental in dealing with complements.

LEMMA 6. *Let L be a complement, and $I(L)$ the inversion of the family L. The family $R(L)$ of all nonclosed partial solutions in $I(L)$ is a direct complement.*

PROOF. By the definition of a dual complement, there exist a closed family D and a family $G \subseteq B$ such that the structure $(LG \mid D)$ is dual. By Lemma 4, so is the structure $(LB|D)$. Consider the identical interpretation φ of the template $(LB \mid D)$: $\varphi(LB) \equiv LB$ and $\varphi(D) = D$. Both the structure $(B \mid D)$ and the template $(LB \mid D)$ satisfy Condition (∗) of Lemma 3. The role of the family U is played by L. By Lemma 3, the structure $(B \mid DI(L))$ is dual.

If $R(L)$ is the family of all nonclosed partial solutions in $I(L)$, then $D' = I(L) \setminus R(L)$ is a family of all partial solutions and DD' is a closed family. Thus, we have obtained the dual structure $(B \mid DD'R(L))$, i.e., $R(L)$ is, by definition, a direct complement.

Lemma 6 associates with each dual complement L a direct complement $R(L)$ which we shall call the *inverted complement* to L.

One can define an operation of addition for complements, which will make it possible to make the columns in complements longer. Note that any column of length m (by length is meant the cardinality of the column) necessarily contains a closed partial solution, while any column of length $m + 1$ contains an alternative set. Therefore, no direct complement can include columns of length greater than m, and no dual complement can contain columns of length greater than $m + 1$.

The general rule of addition for complements and templates is given in the appendix. We shall now describe the simplest type of *addition rule*—the *branching rule*.

Let S_1 and $S_2 = (t_1, t_2, \ldots, t_k)$ be two families of columns and l be a column in S_1. We shall say that the family

$$\begin{pmatrix} l & l & & l \\ t_1 & t_2 & \ldots & t_k \end{pmatrix}$$

is obtained by *branching the column l over S_2*, and denote the result by $\begin{pmatrix} l \\ S_2 \end{pmatrix}$.

If the family S_1 does not absorb S_2, then there exists a column $l \in S_1$ such that all the columns of the family $\begin{pmatrix} l \\ S_2 \end{pmatrix}$ are longer than the column l.

Indeed, the definition of absorption implies that there exists a column $l \in S_1$ containing no column from S_2. Therefore l is a proper subset of every column in the family $\begin{pmatrix} l \\ S_2 \end{pmatrix}$, as required.

Now let S_1 and $S_2 = (t_1, t_2, \ldots, t_k)$ be complements (both of which are primal or dual simultaneously). We shall show that the family K of all nonclosed columns in $\left((S_1 \setminus \{l\}) \cup \begin{pmatrix} l \\ S_2 \end{pmatrix}\right)$ is also a complement.

For definiteness, let S_1 and S_2 be primal complements. Then there exist dual structures $(G_1 \mid S_1 D_1)$ and $(G_2 \mid S_2 D_2)$. It is easy to check that the structure

$$(G_1 G_2 \mid S_1 \setminus \{l\} \begin{pmatrix} l \\ S_2 \end{pmatrix} D_1 D_2)$$

is also dual: any path from the right passes either through $S_2 D_2$ (and then contains a column from G_2), or through $S_1 D_1$ (and then contains a column from G_1). Therefore, the family $K \subset S_1 \setminus \{l\} \begin{pmatrix} l \\ S_2 \end{pmatrix}$ satisfies the definition of a primal complement ($G_1 G_2 \subset B$, $D_1 D_2$ is a closed family).

Finally, let us mention another useful way of obtaining complements.

Let R be a primal complement. Suppose that each partial solution $r \in R$ is completed to a closed partial solution δ_r, and let $c_r = \delta \setminus r$. The set $u = \bigcup_{r \in R} c_r$ will be called the *slack of the complement R*. A set u is said to be a *slack* if it is a slack for some primal complement.

LEMMA 7. *Let u be a slack. Then its transpose u^T is a dual complement.*

PROOF. For definiteness, let u be the slack of the complement R. Since R is a primal complement, there exists a closed family D such that the structure $(B \mid RD)$ is dual. Let $D' = \{\delta_r \mid r \in R\}$. Then the structure $(u^T B \mid D'D)$ is also dual. Indeed, any path in $D'D$, that passes thorough none of the completions $c_r, r \in R$, is a path in RD and contains a column in B, while a path passing through at least one completion contains an element of the slack u, i.e., a column from u^T. Since $D'D$ is a closed family, then u^T satisfies the definition of a dual complement.

Suppose we have obtained a family of slacks U. Each slack $u \in U$ "extends" some primal complement R_u to a family D_u which is closed by some rec_u. Let $\text{rec} = \min\{\text{rec}_u \mid u \in U\}$. We shall say that U is the family of slacks for rec.

LEMMA 8. *Let U be the family of slacks for* rec. *If $(B \mid U)$ is a dual structure, then* rec = opt.

The proof of Lemma 8 is given in the appendix.

COMMENT TO LEMMA 8. The duality of the structure $(B \mid U)$ means that U is a section of the primal tree. Construct the family S^U by branching each slack $u \in U$ over its primal complement R_u. The family S^U is closed by rec. Since each primal complement R_u is by definition accompanied by its closed family D^u, then $D^U = \{D^u \mid u \in U\}$ is a closed family. So is the family $(S^U D^U)$. It is easy to show (using the same argument as for branching complements) that the structure $(B \mid S^U D^U)$ is dual. Thus, the completeness of the family of slacks U is equivalent to the existence of a closed section $S = (S^U D^U)$.

LEMMA 9. *If $(B \mid UU')$ is a dual structure, U a family of slacks, and U' a family of nonclosed partial solutions, then U' is a primal complement.*

The proof consists in constructing the family S^U discussed in the above comment to Lemma 8. The duality of the structure $(B \mid S^U U' D^U)$ is established in a straightforward manner.

The notions introduced above make it possible to construct algorithms differing by the number of complements used at each step, as well as by the rules of processing them. Consider one of possible schemes.

7. A scheme for the algorithm

Initial step. 1. Using heuristic considerations, find a solution β_r with the value of the functional $f(\beta_r) = $ rec and some closed partial solutions forming the closed family D. (See the example in Section 8.)

2. Take an interpretation of the template $(X \mid Y)$ such that $\varphi(Y) = D$. Nonclosed sets from $\varphi(X)$ are those constituting the initial dual complement L. (One can split D into a number of parts and take a separate template for each part.)

3. For the initial primal complement R take the inverted complement $I(L)$. (If D is split into parts, the primal and the dual complements can be obtained from different parts.)

Ordinary step. 1. Form a family U_R of different but not necessarily all slacks of the complement R.

2. Interpret a template $(X \mid Y)$ in such a way that $\varphi(X) \subseteq B$. If $\varphi(Y) \subseteq U_R$, then the problem is solved. Otherwise, for each $u \in U' = \varphi(Y) \setminus U_R$ check whether u is a slack for the complement R; compute est(s) for each $s \in S_u$ and compare est(s) with rec. If every $s \in \bigcup_{u \in U'} S_u$ is closed (est$(s) \geq $ rec), then the problem is solved. Otherwise, form a primal complement R' out of all nonclosed partial solutions $s \in \bigcup_{u \in U'} S_u$. In the case where s is a solution, compute $f(s)$ and set rec $= \min\{f(s), \text{rec}\}$.

Construct a slack $u_{R'}$ of the complement R' such that the transposed family $u_{R'}^T$ is not absorbed by the dual complement L.

3. Extend the column $l \in L$ (or the columns $l \in L$) for which $l \cap u_{R'} = \emptyset$, thus forming the family $l' = \begin{pmatrix} l \\ u_{R'}^T \end{pmatrix}$ by branching l over $u_{R'}^T$. Construct the family $L' = \{L \setminus \{l\}\} \cup l'$.

4. For the new dual complement L take the family L of all nonclosed sets in L'. For the new primal complement R take the inverted complement $I(L)$. If both complements are nonempty, go to the next ordinary step.

The convergence of the algorithm after at most 2^m steps follows from the fact that at each step there is a column in the complement which becomes longer at this step, while the number of different columns in the complement does not exceed 2^m.

For machine realization of the algorithm one has to specify the technique for constructing estimates and slacks, and for selecting templates. We now show how to do this, using as an example the multidimensional knapsack problem.

8. Example

Consider the multidimensional knapsack problem

$$\max \sum_{i \in M} c_i x_i, \qquad M = \{1, \dots, m\},$$

under the conditions

(1e) $$\sum_{i \in M} a_{ij} x_i \leq b_j, \qquad j \in N = \{1, \dots, n\},$$

(2e) $$x_i \in \{0, 1\}, \qquad i \in M,$$

where $a_{ij}, b_i, c_i \geq 0$.

Here every alternative set is of the form $B_i = \{i, \bar{i}\}$, and the family $B = \{B_i \mid i \in M\}$. We shall estimate partial solutions with the use of linear programming. Suppose that we have solved the linear problem

$$\max\left\{\sum_{i \in M} c_i x_i \mid (1e), \quad 0 \leq x_i \leq 1, i \in M\right\}.$$

Let $v_j, j \in N$, be optimal values of the dual variables corresponding to restrictions (1e) in this problem.

Consider a relaxed condition

$$\sum_{i \in M} a_i x_i \leq b,$$

where

$$a_i = \sum_{j \in N} a_{ij} v_j, \qquad b = \sum_{j \in N} b_j v_j.$$

Each partial solution $\delta = \delta[I]$ is defined by the set of indices $I \subset M$ and its partition into $K_0 = \{i \in I \mid \delta[i] = \bar{i}\}$ and $K_1 = \{i \in I \mid \delta[i] = i\}$. Let $K_2 = M \setminus I$. In terms of the variables x_i we have: $x_i = 0$ for $i \in K_0$, $x_i = 1$ for $i \in K_1$, and no value is assigned to x_i for $i \in K_2$. For example, if $m = 7$ and $\delta = \{\bar{2}, 4, \bar{5}, 7\}$, then we have $K_0 = \{2, 5\}$, $K_1 = \{4, 7\}$, $K_3 = \{1, 3, 6\}$.

Solution β is characterized by the empty set K_2. A solution β is said to be admissible, if
$$\sum_{i \in K_1} a_{ij} \leq b_j, \qquad j \in N.$$
If
$$\sum_{i \in K_1} > b_j$$
for at least one $j \in N$, then the solution is inadmissible.

If β is an admissible solution, we set $f(\beta) = \sum_{i \in K_1} c_i$; otherwise $f(\beta) = 0$.

Following [2] we define the estimate of the partial solution δ with the partition K_0, K_1, K_2 as follows:
$$\text{est}(\delta) = \sum_{i \in K_1} c_i + \psi(K_1, K_2),$$
where
$$\psi(K_1, K_2) = \max \sum_{i \in K_2} c_i x_i,$$
$$\sum_{i \in K_2} a_i x_i \leq b - \sum_{i \in K_1} a_i,$$
$$0 \leq x_i \leq 1, \qquad i \in K_2.$$

If this problem has no admissible $x_i, i \in K_2$, then $\psi(K_1, K_2) := 0$. To compute $\text{est}(\delta)$ let us order the ratios c_i/a_i, $i \in K_2$, in such a way that
$$\frac{c_{i_1}}{a_{i_1}} \geq \frac{c_{i_2}}{a_{i_2}} \geq \ldots \geq \frac{c_{i_k}}{a_{i_k}}, \qquad \text{where } k = |K_2|.$$

Determine i_0 for which
$$\sum_{i \leq i_0} a_i \leq b - \sum_{i \in K_1} a_i < \sum a_i,$$
and set $x_i = 1$ for $i < i_0$, and $x_i = 0$ for $i > i_0$;
$$x_{i_0} = b - \sum_{i \in K_1} a_1 - \sum_{i \leq i_0} a_i,$$
i.e.,
$$\text{est}(\delta) = \sum_{i \in K_1} c_i + \sum_{i \leq i_0} c_i + c_{i_0} x_{i_0}.$$

In the specific example considered below we shall assume that the linear problem is already solved, the dual variables are determined, and the ordering is already done.

Since in the problem under consideration we have to find maximum and not minimum, the conditions defining the estimate and the closed partial solution are, respectively, of the form $\text{est}(\delta) \geq f(\beta)$, $\text{est}(\delta) \leq \text{rec}$.

Now let us solve a specific example defined by the parameters $m = 9, n = 2$.

i	1	2	3	4	5	6	7	8	9
c_i	9	19	13	11	8	11	6	3	7
a_{i1}	0	7	3	5	2	4	2	1	4
a_{i2}	6	2	5	1	4	4	4	2	6
a_i	6	16	11	11	8	12	8	4	14
$\frac{c_i}{a_i}$	1.5	1.2	1.2	1	1	0.9	0.7	0.7	0.5

b_1	13
b_2	16
b	42

v_1	2
v_2	1

We shall use two heuristics in order to obtain a closed family D. The first one is to "pick up items with greatest specific cost", the second is to "pick up most expensive items".

The first heuristic gives the solution $\beta_1 = \{1, 2, 3, \bar{4}, 5, \bar{6}, \bar{7}, 8, \bar{9}\}$ with the value of the functional $f(\beta_1) = 44$, while the second yields $\beta_2 = \{\bar{1}, 2, 3, \bar{4}, 5, \bar{6}, \bar{7}, 8, \bar{9}\}$ with the value of the functional $f(\beta_2) = 43$. The record is $\text{rec} = 44 = \max\{f(\beta_1), f(\beta_2)\}$. Since for the partial solution $\delta_1 = \{1, 2, 3\}$ the only admissible extension is the solution β_1, and for $\delta_2 = \{2, 3, 5\}$ the only admissible extension is β_2, then $\text{est}(\delta_1) = 44, \text{est}(\delta_2) = 43$. Thus we have constructed a family $D = (\delta_1 \delta_2)$ which is closed by the record $\text{rec} = 44$. We shall use the template (f) in order to construct the primal complement R, and (e) to construct the dual complement. The template (f) is of the form

$$(X \mid Y) = \begin{pmatrix} a & c & e & \begin{array}{cc} a & b \\ e & g \end{array} & b & d & a \\ b & d & g & \begin{array}{cc} c & d \end{array} & c & e & g \end{pmatrix}.$$

Consider an interpretation φ for which $\varphi(X) \subset B$ and one of the columns in $\varphi(Y)$ is δ_1. With this in mind, set $\varphi(a) = 2, \varphi(b) = \bar{2}, \varphi(c) = 1, \varphi(d) = \bar{1}, \varphi(e) = 3, \varphi(g) = \bar{3}$. We obtain the dual structure

$$\begin{pmatrix} 2 & 1 & 3 & \begin{array}{cc} 2 & \bar{2} \\ 3 & \bar{3} \end{array} & 2 & \bar{1} & 2 \\ \bar{2} & \bar{1} & \bar{3} & \begin{array}{cc} 1 & \bar{1} \end{array} & 1 & 3 & \bar{3} \end{pmatrix}.$$

We have to provide estimates for the four new partial solutions $\delta_3 = \{\bar{1}, \bar{2}, \bar{3}\}, \delta_4 = \{\bar{2}, 1\}, \delta_5 = \{\bar{1}, 3\}, \delta_6 = \{2, \bar{3}\}$. The calculations yield $\text{est}(\delta_3) = 38.3, \text{est}(\delta_4) = 46.5, \text{est}(\delta_5) = 47, \text{est}(\delta_6) = 49$. Since $\text{est}(\delta_j) > 44$ for $j = 4, 5, 6$, then the partial solutions δ_j, $j = 4, 5, 6$, are nonclosed, and, by definition, the family

$$R = \begin{pmatrix} \bar{2} & \bar{1} & 2 \\ 1 & 3 & \bar{3} \end{pmatrix}$$

is a primal complement.

In order to construct the dual complement, consider an interpretation φ of the template (e)

$$(X \mid Y) = \begin{pmatrix} a & a & b & b & c & c & d & \begin{array}{c} a \\ b \end{array} & d & f \\ d & e & e & f & f & g & g & c & e & g \end{pmatrix}$$

for which one of the columns in $\varphi(Y)$ is the partial solution δ_2, and $\varphi(X)$ contains as many alternative sets as possible. With the in mind, set $\varphi(a) = 2, \varphi(b) = 3, \varphi(c) = 5, \varphi(d) = \bar{2}, \varphi(e) = \bar{3}, \varphi(f) = \bar{3}, \varphi(g) = \bar{5}$. We obtain the dual structure

$$\begin{pmatrix} 2 & 2 & 3 & 3 & 5 & 5 & \bar{2} & 2 & 3 & \bar{2} & \bar{3} \\ \bar{2} & 3 & \bar{3} & 3 & \bar{5} & 5 & \bar{5} & 5 & \bar{3} & \bar{5} \end{pmatrix}.$$

Let us estimate the partial solutions $\delta_7 = \{\bar{2}, \bar{3}\}$ and $\delta_8 = \{\bar{3}, \bar{5}\}$. We have $\text{est}(\delta_7) = 39.7 < \text{rec}$ and $\text{est}(\delta_8) = 43.6 < \text{rec}$, i.e., δ_7 and δ_8 are closed partial solutions. By definition, the family of nonclosed sets

$$L = \begin{pmatrix} 2 & 5 & \bar{2} \\ 3 & 3 & \bar{5} \end{pmatrix}$$

is a dual complement.

The initial step is completed.

Let us now perform the ordinary step. Construct the slack for the complement R. Choose the partial solution in R having the maximal estimate, i.e., $\delta_6 = \{2, \bar{3}\}$. Complete this solution with the alternatives 5 and 1. (If it turns out that $u = \{1, 5\}$ is a slack, then the dual complement u^T is not absorbed by the complement L. It is this property that one should follow in selecting the slack.) The partial solution $\delta_9 = \{2, \bar{3}, 5, 1\}$ has the following characteristics:

$$\sum_{i \in K_1} a_{i1} = 7 + 2 + 0 = 9; \quad b_1 - 9 = 4;$$

$$\sum_{i \in K_1} a_{i2} = 2 + 4 + 6 = 12; \quad b_2 - 12 = 4.$$

Of the remaining indices $i = 4, 6, 7, 8, 9$ only one of $i = 6, 7, 8$ "fits". Choosing the most expensive index $i = 6$, and adding it to δ_9, we obtain the new solution $\beta_3 = \{2, \bar{3}, 5, 1, 6, \bar{4}, \bar{7}, \bar{8}, \bar{9}\}$ with the value of the functional $f(\beta_3) = 47$. This modifies the record value $\text{rec} = \max\{\text{rec}, f(\beta_3)\} = 47$. As an estimate (with respect to the best completion) of the partial solution δ_9 one can take $\text{est}(\delta_9) = 47$. Completing the partial solutions δ_1 and δ_5 by the alternative 5, we obtain partial solutions $\delta_{10} = \{\bar{2}, 1, 5\}$ and $\delta_{11} = \{\bar{1}, 3, 5\}$ with the estimates $\text{est}(\delta_{10}) = 46.5 < 47$ and $\text{est}(\delta_{11}) = 47 = \text{rec}$. Thus, the set $u = \{1, 5\}$ is the slack for the complement R. Since u^T is not absorbed by the complement L, we go directly to step 3.

Form the family

$$L' = \begin{pmatrix} 2 & 2 & 5 & \bar{2} & \bar{2} \\ 3 & 3 & 3 & \bar{5} & \bar{5} \\ 5 & 1 & & 5 & 1 \end{pmatrix}$$

by branching the columns $\{2, \bar{3}\}$ and $\{\bar{2}, \bar{5}\}$ from L over u^T. Note that the first column in L' is absorbed by the third, while the fourth column is closed. Therefore, the new dual complement is

$$L_1 = \begin{pmatrix} 2 & 5 & \bar{2} \\ 3 & 3 & \bar{5} \\ 1 & & 1 \end{pmatrix}.$$

To construct the new direct complement we consider the inverse

$$I(L_1) = \begin{pmatrix} \bar{2} & \bar{5} & 2 \\ 3 & 3 & 5 \\ \bar{1} & & \bar{1} \end{pmatrix}.$$

Estimating the partial solutions $\delta_{12} = \{\bar{2}, 3, \bar{1}\}, \delta_{13} = \{\bar{5}, 3\}, \delta_{14} = \{2, 5, \bar{1}\}$, we obtain $\mathrm{est}(\delta_{12}) = 43 < \mathrm{rec}, \mathrm{est}(\delta_{13}) = 50 > \mathrm{rec}, \mathrm{est}(\delta_14) = 47 = \mathrm{rec}$. The only nonclosed solution is δ_{13}; hence the primal complement $R_1 = \begin{pmatrix} \bar{5} \\ 3 \end{pmatrix}$. Complete δ_{13} with the alternatives 1 and 2 to the closed partial solution $\delta_{15} = \{\bar{5}, 3, 1, 2\}$ (the fact that it is closed follows from the same property of δ_1). Thus $u = \{1, 2\}$ is a slack. Consider the template

$$\begin{pmatrix} 1 & 2 & \bar{2} & 1 & \bar{1} \\ \bar{1} & 2 & & 2 & 2 \end{pmatrix}.$$

If it turns out that $\{\bar{1}, 2\}$ and $\{\bar{2}\}$ are discrepancies, then the problem is solved (the complete family of discrepancies is constructed). Estimating $\delta_{16} = \{\bar{5}, 3, \bar{1}, 2\}$ and $\delta_{17} = \{\bar{5}, 3, \bar{2}\}$, we obtain $\mathrm{est}(\delta_{16}) = 46.6 < \mathrm{rec}$ and $\mathrm{est}(\delta_{17}) = 45.5 < \mathrm{rec}$, i.e., δ_{16} and δ_{17} are closed, $\{\bar{1}, 2\}$ and $\{\bar{2}\}$ are discrepancies, and the problem is solved.

9. Appendix

9.1. Addition with respect to a template. Let $S_1, S_2, \ldots, S_n$ be finite families of finite sets such that

$$S_j = \{s_j^t \mid t \in K = \{1, \ldots, k\}\}, \quad j \in N = \{i, \ldots, n\},$$

and let $(X \mid Y)$ be a template for which $X = \{x_j \mid j \in N\}, Y = \{y_i \mid i \in L = \{1, \ldots, l\}\}$ and $x_j = \{x_j^t \mid t \in \{1, \ldots, k_j\}\}, j \in N$. For each $i \in N$ set

$$p_i = \bigcup_{\{t, j \mid x_j^t \in y_i\}} s_j^t.$$

We shall say that the family $P = \{p_i \mid i \in L\}$ is obtained by adding the families $S_j, j \in N$, with respect to the template $(X \mid Y)$, and write

$$P \stackrel{(X \mid Y)}{=} \sum_{j \in N} S_j.$$

LEMMA 1A. *Let $P \stackrel{(X \mid Y)}{=} \sum_{j \in N} S_j$. Then any path in the family P contains a path in some family $S_j, j \in N$.*

PROOF. Let π be a path in P, i.e., $\pi = \bigcup_{i \in L} \pi_i$, where $\pi_i \in p_i$. By the definition of the set p_i the element π_i belongs to at least one set s_j^t such that $x_j^t \in y_i$. Choose one of such sets $s_{j(i)}^{t(i)}$. Since $(X \mid Y)$ is a template, the path $\{x_{j(i)}^{t(i)} \mid i \in L\}$ in Y contains a column x_j from the family X, and the corresponding family of sets $G = \{s_{j(i)}^{t(i)} \mid i \in L\}$ contains the family S_J. Since π is a path in G, π contains a path in S_j. The lemma is proved.

LEMMA 2A. *Let* $(B_1 \mid S_1D_1), (B_2 \mid S_2D_2), \ldots, (B_n \mid S_nD_n)$ *be templates and* $P \stackrel{(X|Y)}{=} \sum_{j \in N} S_j$. *Then the structure* $(B \mid D) = (B_1B_1 \ldots B_n \mid PD_1 \ldots D_n)$ *is dual.*

PROOF. Consider any path π in D. Let $\pi(P) = \pi \cap P$, i.e. $\pi(P)$ is a path in P. By Lemma 1A there is an index $j \in N$ such that $\pi(P)$ contains a path $\pi(S_j)$ in the family S_j. Selecting in the original path π the subpath $\pi(S_jD_j) = \pi(S_j) \cup \pi(D_j)$, we construct $\pi(S_jD_j) = \pi(S_j) \cup \pi(D_j)$, which is a path in the family S_jD_j. Since $(B_j \mid S_jD_j)$ is a template, then $\pi(S_jD_j)$ has a nonempty intersection with any path $\rho(B_j)$ in B_j. But any path ρ in B contains a path $\rho(B_j)$, and it follows from $\pi(S_jD_j) \cap \rho(B_i) \neq \emptyset$ that $\pi \cap \rho \neq \emptyset$. The lemma is proved.

Lemma 2A makes it possible, for empty D_j, $j \in N$, to generate new templates from those already constructed. Nonempty D_j can be considered as families in the definitions of primal and dual complements. This allows one to use the addition of complements with respect to templates.

Let $K_1, K_2, \ldots, K_n$ be complements (all of them either primal or dual simultaneously), $(X \mid Y)$ a template, and $P \stackrel{(X|Y)}{=} \sum_{j \in N} K_j$. The family K of all nonclosed columns in P is a new complement. The fact that K is a complement follows from Lemma 2A.

9.2. Proof of Lemma 8. We must prove that if U is the family of discrepancies for rec and $(B \mid D)$ is a dual structure, then rec = opt.

For each slack $u \in U$ form the transposed family u^T, i.e., the dual complement. Sum up all $u^T, u \in U$, with respect to the template $(U \mid B)$. The family $P \stackrel{(U|B)}{=} \sum_{u \in U} u^T$ coincides with the family $B = \{B_i \mid i \in M\}$, in which all the sets are alternative, i.e. closed. Therefore, the family K of nonclosed columns in B is empty. We have thus obtained an empty dual complement K, which, as mentioned in the definition of a complement, means that rec = opt.

9.3. A version of the algorithm. Using the addition of complements, one can construct algorithms of the following type. The objective of the algorithm is to obtain an empty complement. The parameter of the algorithm is the number k of the complements under consideration.

The *initial step* consists in constructing a stock S of trivial dual complements and k dual complements L_j whose columns, in general, do not consist of a single element. For example, if D is a closed family obtained heuristically, then S consists of all transposed partial solutions $\delta^T, \delta \in D$. In order to obtain k complements, we select k templates $(X_j \mid Y_j)$, $j \in K = \{1, \ldots, k\}$, and their interpretations $\varphi_j, j \in K$, such that $\varphi_j(Y_j) \subseteq D$ and each family $\varphi_j(X_j)$ contains alternative sets. The complement $L_j, j \in K$, is the family of nonclosed columns in $\varphi_j(X_j)$.

The *ordinary step* consists in defining the discrepancies for the primal complements $R(L_j)$ from the inverted dual complements $I(L_j)$, $j \in K$, in such a way that either the entire family U of the discrepancies thus obtained is dual to $B = \{B_i \mid i \in M\}$, or the addition of some complement $u^T, u \in U$, to the dual complement $L_j, j \in K$, makes the columns of the latter longer.

All nonabsorbed trivial complements, including $u^T, u \in U$, form the new stock S. If one of the complements $L_j, j \in K$, becomes empty after lengthening, the algorithm is terminated; otherwise go to the new ordinary step. The record is recomputed in the process of defining discrepancies for primal complements.

In the case $k = 1$ the dual complement L can be lengthened without discrepancies. Continue the nonclosed partial solution δ of the inverted complement $R(L)$ to a closed partial solution δ_r such that $\delta_r^T \cap I(\delta) = \emptyset$. Consider the family $\begin{pmatrix} I(\delta) \\ \delta_r^T \end{pmatrix}$ obtained by branching $I(\delta)$ over δ_r^T. All the sets in this family are closed, except for those having the form $I(\delta) \cup \{a\}$, where $a \in \delta_r \setminus \delta$. Eliminating from these sets the sets l for which $I(l)$ are closed partial solutions, and going back to the primal complement $R(L)$, we see that the partial solution δ is branched over some alternatives of the set $\delta_r \setminus \delta$. This models one step in the branch-and-bound method.

9.4. Complements in general case (alternative sets, generally speaking, are not 2-element sets). We shall describe the reduction to the case of 2-element alternative sets. Along with the alphabet $A = \bigcup_{i \in M} B_i$ introduce the alphabet A' such that each alternative $a \in A$ is associated with the literal $a' \in A'$. Setting $n = |A|$, form n 2-element sets $H_a = \{a, a'\}$, $a \in A$, and the family $H = \{H_a \mid a \in A\}$. If X is a family in the alphabet A, then X' will denote the family obtained from X by replacing each alternative a with a'.

LEMMA 3A. *The structure $(B \mid D)$ is dual in the alphabet A if and only if the structure $(H \mid BD')$ is dual in the alphabet $A \cup A'$.*

PROOF. Let $(B \mid D)$ be a dual structure. Then any paths in B and D have nonempty intersection. This means that any path in BD' contains both members of some alternative a and a', i.e., contains an alternative set H_a, and so the structure $(H \mid BD')$ is a dual one.

On the other hand, any paths π in B and ρ in D form the path $\pi \cup \rho'$ in BD'. If the structure $(H \mid BD')$ is dual, then $\pi \cup \rho'$ contains an alternative set H_a, and therefore the paths π and ρ have a as their intersection, so that $(B \mid D)$ is a dual structure.

Lemma 3A makes it possible to construct sections of the tree in which branchings are defined by alternative sets in the family H. Let us agree to use the alphabet A' to write partial solutions in the original problem, and the alphabet A to write alternative sets B_i. The set $\alpha \in A \cup A'$ will be called closed if it contains either a closed partial solution in the alphabet A', or an alternative set in the alphabet A.

The families $H_a^T = (aa')$ are trivial primal complements. When summed up with respect to the template, they result in families of sets containing alternatives from both alphabets A and A'. After eliminating all closed sets from such a family, we obtain a primal complement.

In a dual tree, branchings are defined both by alternative sets $B_i, i \in M$, and by partial solutions δ' from the closed family D'. Thus the trivial dual complements are both $B_i^T, i \in M$, and $\delta'^T, \delta' \in D'$. When summed up with respect to the template, trivial dual complements result in families of columns. Such a column is closed if it contains a set $H_a, a \in A$. After eliminating all closed columns from the family, one obtains a dual complement.

Taking for the closed set the union of the closed set D' and the family $B = \{B_i \mid i \in M\}$, we reduce the general problem to the problem with 2-element alternative sets.

10. Conclusion

The technique proposed in this article makes it possible to construct sections of arbitrary form, which in turn makes it possible to consider sections of partial solutions that are sufficiently remote from one another in any solution tree.

Making use of arbitrary sections instead of a trivial tree form, we do not have to worry about ordering tree levels, because a section retains this property under any ordering of indices $i \in M$. A successfully chosen heuristic must only define the family of active indices $i \in M$ participating in the section.

In order to construct short closed partial solutions one must have, as in the branch-and-bound method, a technique for obtaining powerful estimates aimed at the special form of the partial solution. For this purpose one may use submodular objective functions [7].

Dual structures can be used not only to solve but also to formulate problems. Suppose that an analysis of specific conditions yields a closed family D. Then the dual structure $(B \mid D)$ defines a family B of alternative sets describing the problem for which rec = opt. On the other hand, using the family of alternative sets B and the dual structure $(B \mid D)$, we obtain the family D; the fact that it is closed under an appropriate choice of the functional means that the record in the problem described by the alternative sets of the family B is optimal.

We are grateful to G. E. Mints for his interest in the work and useful comments.

References

1. G. V. Davydov and I. M. Davydova, *Duality in search problems*, Third Conf. "Application of Methods of Mathematical Logic", Abstracts of Reports, Tallinn, 1983, pp. 73–74. (Russian)
2. I. V. Romanovsky, *Algorithms for solving extremal problems*, "Nauka", Moscow, 1977. (Russian)
3. I. M. Davydova, *Methods for solving positioning problems*, Operations Analysis and Statistical Modelling, vol. 4, Leningrad, 1977, pp. 50–80. (Russian)
4. G. V. Davydov, *The synthesis of the resolution method and the inverse method*, Zap. Nauchn. Sem. Leningrad. Otdel. Mat. Inst. Steklov. (LOMI) **20** (1971), 24–35; English transl., J. Soviet Math. **1** (1973), 12–18.
5. A. D. Gvishiani and V. A. Gurvich, *Dual systems of sets and their applications*, Izv. Akad. Nauk SSSR Tekhn. Kibernet. **1983**, no. 4, 31–39; English transl., Engrg. Cybernetics **21** (1983), no. 4, 112–118.
6. N. K. Zamov and V. I. Sharonov, *Application of programs in deduction search*, Studies in Applied Mathematics, vol. 1, Kazan, 1973, pp. 94–102. (Russian)
7. L. Lovász, *Submodular functions and convexity*, Mathematical Programming: the State of the Art (Bonn, 1982), Springer-Verlag, Berlin, 1983, pp. 235–257.

Models, Methods, and Modes for the Synthesis of Program Schemes

A. Ya. Dikovsky

ABSTRACT. The author describes and illustrates by multiple examples various formulations of the problem of automated synthesis of linear, branching, and recursive programs on the basis of functional and operator dependencies leading to fast synthesis algorithms. A survey of complexity estimates for these algorithms is given.

1. Introduction

A special feature of applied computations is that variables therein play the roles of both theoretical notions and problem parameters, and, therefore, possess their own semantics. In this respect they are different from, say, program variables that are essentially names enabling either direct or indirect access to values. For example, in plane geometry the following variables are used to state problems about triangles: angles $A, B, C,$, lengths of the sides a, b, c, lengths of the corresponding heights h_a, h_b, h_c, the area S, the semiperimeter p, the radii of the incircle and the circumcircle r and R, and other variables. Similarly, in accounting one uses *type of payment, type of deduction, actual time, amount to be paid*, etc. The list of such examples is inexhaustible. Relations connecting variable values constitute a theory for a given class of applied problems, and more often than not they include functions expressing values of one variable via other variables. Thus in an example with triangles, the value of S can be determined by various functions:

$$S = \frac{1}{2}h_a a = \frac{1}{2}h_b b = \frac{1}{2}h_c c = \frac{1}{2}bc \sin A$$
$$= \frac{a^2 \sin B \sin C}{\sin A} = \sqrt{p(p-a)(p-b)(p-c)}.$$

In many applied problems the task is to express desired values in a corresponding applied theory through other variables whose values in the problem are assumed to be known. A solution for such a computational problem, if it exists, is given by a superposition of functions derived from the relations of the theory. In our example with triangles the task may be to find the area given the sides of the triangle. Using

©1996 American Mathematical Society

Heron's formulas and the relation $p = \frac{1}{2}(a+b+c)$, we obtain the expression

$$S = \sqrt{\frac{1}{2}(a+b+c)\left(\frac{1}{2}(a+b+c) - a\right)\left(\frac{1}{2}(a+b+c) - b\right)\left(\frac{1}{2}(a+b+c) - c\right)}.$$

Essentially, this expression describes a program computing the value of the parameter S through the values of the parameters a, b, c. However, it is not difficult to see that in order to prove the existence of a solution there is no need to know how the functions involved in it are computed. It is sufficient to know the names of the functions, the variables whose values they compute, the variables they depend on, and conditions under which these functions are defined. In other words, in order to find a solution, it is sufficient to consider functional schemes (also called dependencies) of the form $f: R \to (A_1 \ldots A_k \to A)$. The meaning of such a dependency is that under the condition R we have $A = f(A_1 \ldots A_k)$. Thus, in our example with triangles, we can specify the following dependencies

$$\sigma_1: h_a a \to S,$$
$$\sigma_2: Abc \to S,$$
$$\sigma_3: aABC \to S,$$
$$\sigma_4: pabc \to S,$$
$$\pi: abc \to p,$$
etc.

Now, if in the above program for computing the area we replace the functions whose superposition computes the area by their names, we obtain the program scheme $S = \sigma_4(\pi(a,b,c), a, b, c)$. The program itself is obtained from this scheme through an appropriate interpretation of the names. Under a program realization the interpreting functions are realized by program modules found in applied libraries.

Applied theories formulated as sets of dependencies were called computational models by E. H. Tyugu in [1]. In this paper, synthesis in computational models is understood as searching for and constructing program schemes solving computational problems. Although different semantics, either operator or deductive, are used to interpret computational models, all of them are constructive in the sense of Brouwer (see [2]) which was made more precise by Kolmogorov [3], Kleene [4], and other logicians: a proof of existence for a solution of a computational problem is essentially equivalent to constructing a solution. Furthermore, in computational models of known types the program scheme solving a computational problem is obtained from the proof of its existence in linear time with respect to the written length of the proof. Thus, in these models the problem of effective search for a solution (the synthesis problem) is reduced to the problem of effective search for the proof of its existence (the analysis problem).

In this paper we briefly review the classification of computational models by the types of dependencies. First by way of examples, and then in precise form, we describe various forms and solution modes for computational problems. The main distinction we make is between partial solutions, which in this work are represented by terms, and general solutions obtained by interpreting computational models via recursive program schemes. For each type of computational model and for all modes and types of solutions of computational problems we formulate the analysis and synthesis problems, as well as some related algorithmic problems. We give no descriptions of effective algorithms solving the algorithmic problems, because this would have significantly increased the length of the paper. Instead we describe the solution methods on which the algorithms are based, and refer the reader to the papers containing complete descriptions of the algorithms.

2. Solution types and modes for computational problems

Let $\{A_1, A_2, \ldots, A_k, \ldots\}$ be the set of all variables, $\{D_1, D_2, \ldots, D_m, \ldots\}$ the set of dependency names, and $\{R_1, R_2, \ldots, R_n, \ldots\}$ the set of condition names.

DEFINITION 2.1. An *unconditional functional dependency* (FD) is an expression of the form $(W \to V)$, in which W and Y are (possibly empty) collections (i.e., finite linearly ordered sets) of variables. The collection $\{A_1, \ldots, A_r\}$ will, for brevity, be written as a word $A_1 \ldots A_2$. An *unconditional operator dependency* (OD) is an expression of the form

$$\underset{1}{\overset{m}{\&}}\, \alpha_j \to \underset{1}{\overset{n}{\&}}\, \beta_j,$$

in which $m, n \geq 0$ and α_i, β_i are either collections of variables, or FD, or OD (in the last two cases the α_i are called *subproblems*). An expression of the form $R(W)$, where W is a collection of variables, will be called a *relation*. A *conditional functional dependency* (CFD) (respectively, *conditional operator dependency* COD) is an expression of the form $Q \to \alpha$, in which α is an FD (respectively, OD) and Q is the name of a condition or relation. An expression of the form $D_i\!:\!\alpha$, where α is a (possibly conditional) FD or OD, will be called a *named* FD (respectively, OD, CFD, or COD).

A *computational model* is any finite set of dependencies Σ. Let us agree that in named dependencies of the form $D_i\!:\!\alpha$ of the model Σ the name of D_i uniquely defines α. A computation model Σ is said to be an FD *model*, OD *model*, CFD *model*, COD *model*, if Σ contains, respectively, only FD; only FD and OD; only FD and CFD; or any dependencies of the types listed above.

A *computational problem* in a computational model Σ is a system $\Pi = (\Sigma, X, Z)$, in which X and Z are collections of variables. The variables in the collection X are called *initial*, and those in Z are called *goal* variables.

EXAMPLE 2.1. $TRIANGLE_1 = \{\sigma_1, \sigma_2, \sigma_3, \sigma_4, \pi\}$, the fragment of the general computational model $TRIANGLE$ described in the introduction, is itself an FD model. In this model one can formulate, for example, the computational problem

$$(TRIANGLE_1; abc, S),$$

whose solution scheme was described in the introduction in the form of an algebraic expression.

It is convenient to represent FD models and computational models in FD models in the form of bipartite graphs in which variables are vertices of type one, FD

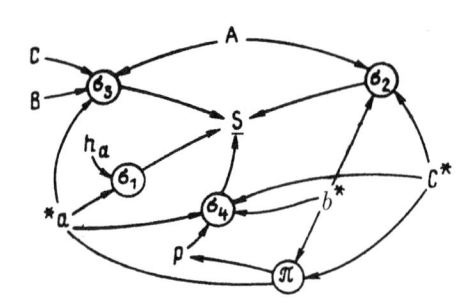

FIGURE 2.1

names are vertices of type two, while FD themselves correspond to arcs. For examples, the computational problem ($TRIANGLE_1, abc, s$) is represented by the graph shown in Figure 2.1

In this graph the vertices corresponding to initial variables are marked with ∗, and the vertex corresponding to the goal variable is underlined. Problem representation in models of other types is similar.

Operator dependencies for which subproblems are not FD rarely arise in practical applications. Furthermore, under reasonable modification of the notion of computational problem, we can restrict our attention to FD of the form

$$f: (W \to A)$$

and OD of the form

$$F: \overset{m}{\underset{1}{\&}} (W_i \to A_i) \to (W \to A),$$

where W, W_i are collections of variables, and A_i, A are variables. In this article we shall consider only such dependencies, since this restriction does not change any essential features of the problem, but substantially shortens the exposition. In specific examples, we shall allow FD of the form $W \to A_1 \ldots A_k$ as an abbreviated notation for the set of FD $W \to A_1, \ldots, W \to A_k$.

EXAMPLE 2.2. With some simplifications, the rule of calculating the amount of advance payment can be described as a computational problem in the model *ADVANCE* with the dependencies

$$f_1: N\ RN \to P\ LE\ SX\ A\ M\ NC\ MS,$$
$$f_2: P\ LE \to SA,$$
$$f_3: SA \to IT,$$
$$f_4: SA\ SX\ A\ M\ NC\ MS,$$
$$f_5: SA\ IT\ UST \to S,$$
$$F: (RN\ SA \to \%) \to (RN \to UST),$$

where the meaning of the variables is as follows:

N — employee's name,
RN — employee's registration number,
P — position,
LE — length of employment,

SX — sex,
M — marital status,
A — age,
NC — number of children,
MS — husband's status with respect to military service,
SA — salary,
IT — income tax,
UST — tax on unmarried and single citizens,
% — part to be deducted,
S — amount to be paid.
The computational problem is of the form

$$(ADVANCE; N\ RN,\ S).$$

In order to solve this computational problem, we first have to solve the subproblem $(RN\ SA \to \%)$. It is not hard to see, however, that this problem cannot be solved precisely in the model $ADVANCE$. Nevertheless, there exists a reasonable strategy for solving the problem.

1. First, by a consecutive application of the dependencies f_1, f_2, and f_3 compute the value of % from N and RN. Since N is given as an initial value, this provides a solution for the subproblem $(RN\ SA \to \%)$.
2. Next, applying the OD F, construct the FD $(RN \to UST)$.
3. Finally, by a consecutive application of the FD $(RN \to UST)$, f_3, and f_5, obtain the value S.

A special feature of this strategy for solving the problem is that the initial variables N and RN serve as a common memory for both the problem and the subproblem.

Add the following FD to the $ADVANCE$ model:

$$f_6: RN \to N.$$

In the new model $ADVANCE_1$ the subproblem $(RN\ SA \to \%)$ becomes solvable, i.e., both the problem and the subproblem become solvable, each under its own conditions (without common memory). For more detail on this, see [5].

Consider another example, taken from [6].

EXAMPLE 2.3. The task of computing the iterated integral

$$\int_0^a \left(\int_0^b z(x,y)\, dx \right) dy$$

can be represented in the form of the computational problem $(\Sigma_{2\text{-int}}; ab, s)$ in the model $\Sigma_{2\text{-int}}$ with the dependencies

$$f_1: xy \to z, \qquad G_1: (x \to z)\ \&\ b \to w, \qquad G_2: (y \to w)\ \&\ a \to s,$$

where the variable w corresponds to the inner integral. It is not hard to see that neither subproblem is solvable in the model $\Sigma_{2\text{-int}}$ even with the common memory ab. Nevertheless, this problem also has a solution strategy:
1. First, one has to solve the subproblem $(x \to y)$ in the framework of the subproblem $(y \to w)$ (i.e., for a given y). Under this condition the subproblem $(x \to z)$ is solved by an application of the FD f_1.
2. Next, using the common memory b and applying the OD G_1, solve the outer subproblem $(y \to w)$.
3. Finally, solve the main problem by applying the OD G_2.

Thus the iterated integration problem cannot be solved by separating the subproblems, i.e., solving each of them under its own conditions (without common memory), or even under the conditions of the main problem (with common memory). This analysis of the solution of the iterated integration problem is given in [5].

Examples 2.1—2.3 demonstrate that formal definitions of a solution to a computational problem in an OD model must reflect various solution modes (with separable or nonseparable subproblems, and with or without common memory).

Let us turn to another aspect of solving computational problems. Enlarge the computational model $TRIANGLE_1$ expressing sides through area and heights, and heights through sides and angles, i.e., adding the dependencies

$$\alpha_1: Sh_b \to b, \quad \alpha_2: Sh_c \to c, \quad \beta_1: cA \to h_b, \quad \beta_2: bA \to h_c.$$

In the new computational model $TRIANGLE_2$ the problem $(TRIANGLE_2; Abc, S)$ has infinitely many solutions

$$\sigma_2(A,b,c), \quad \sigma_2(A, \alpha_1(\beta_1(c,A), \sigma_2(A,b,c)), \alpha_2(\beta_2(b,a), \sigma_2(A,b,c))),$$

etc.

However, under the standard interpretation of the FD of this model all these solutions are equivalent, i.e., define the same values of S for the same values of A, b, c. This property of invariance of a computational problem (under some "standard" interpretation) makes it possible to regard any satisfying expression as its solution. However, this property is not necessarily satisfied in all cases. For example, it does not hold for models with conditional dependencies.

EXAMPLE 2.4. Replace the FD f_3 in the OD model $ADVANCE$ by the following three conditional dependencies:

$$f_4: C_1(SX, A, MS, NC, MS) \to (SA \to \%),$$
$$g_4: C_2(SX, A, MS, NC, MS) \to (SA \to \%),$$
$$h_4: C_3(SX, A, MS, NC, MS) \to \%,$$

where the meaning of the relations C_1, C_2, C_3 under the standard interpretation are

$$C_1(SX, A, M, NC, MS) \rightleftharpoons (SX = \text{"male"}) \ \& \ (A \geq 18),$$
$$C_2(SX, A, M, NC, MS) \rightleftharpoons (SX = \text{"female"}) \ \& \ (A \leq 50)$$
$$\& \ (M = \text{"married"}) \ \& \ (NC = 0) \ \& \ (MS = \text{"no"}),$$
$$C_3(SX, A, M, NC, MS) \rightleftharpoons \neg C_1 \ \& \ \neg C_2.$$

In the new model $ADVANCE_2$, consider the same computational problem

$$(ADVANCE_2; N, RN, S).$$

Then the problem acquires three different solutions, which we shall describe in the form of "pseudoprograms" of the kind used in Examples 2.2 and 2.3:
   solution 1: $f_1; f_2$; if $C_1(SX, A, M, NC, MS)$ then f_4; F; f_3, f_5,
   solution 2: $f_1; f_2$; if $C_1(SX, A, M, NC, MS)$ then g_4; F; f_3, f_5,
   solution 3: $f_1; f_2$; if $C_3(SX, A, M, NC, MS)$ then h_4; F; f_3, f_5.

None of these solutions is, of course, preferable to the others. One could attempt to eliminate this irregularity by trying to find a general solution in the form $f_1; f_2$; if $C_1(SX, A, M, NC, MS) \to f_4 \square C_2(SX, A, M, NC, MS) \to g_4 \square C_3(SX, A, M, NC, MS) \to h_4$ fi; F, f_3, f_5. However, it can be shown that the problem of finding solutions to computational problems in such a form is NP-complete.

A solution to the computational problem in the form of a single satisfying expression will be called a partial solution. The anomalous behavior of partial solutions to noninvariant computational problems becomes even more pronounced if the computational model admits recursion.

EXAMPLE 2.5. Consider the computational model $SORT$ with dependencies

$$\begin{aligned} &\text{ins}: SF \to S, \\ &e_1: R \to S, \\ &\text{cdr}: R \to R, \\ &e_2: M \to R, \\ &\text{car}_1: R \to R, \\ &\text{car}_2: M \to F. \end{aligned}$$

Consider in this model the computational problem

$$\Pi = (SORT; M, S)$$

and choose the standard interpretation under which car_1 and car_2 outputs the first element of the numerical array; cdr deletes the first element from the array; ins inserts the number (the second argument) into the numerical array (the first argument) in such a way that it is preceded only by the elements of the array that are smaller than the inserted number; e_1 and e_2 are identity mappings of the arrays. It is not hard to see that for each n the expression E_n shown in Figure 2.2 in tree form defines a sorting M of every array of length n.

None of these partial solutions E_n is, of course, satisfactory. In addition, there exist infinitely many meaningless partial solutions of the problem Π. It would be desirable to obtain, under interpretation I, a solution of this problem in the form of a relation (or a function)

$$\langle m, s \rangle \in \Pi_{|I|} \equiv E_{|m|}(m) = S.$$

We shall call such a solution of the computational problem a general one.

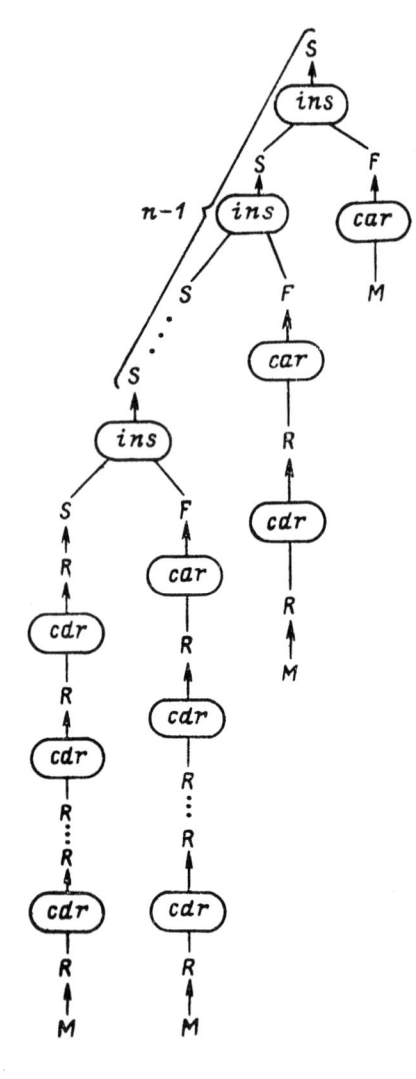

FIGURE 2.2

3. Search for partial solutions

In this section we give precise formulations of the problems of analysis and synthesis arising in the search for partial solutions of computational problems in different modes, and we formulate and discuss algorithmic problems related to the search for such solutions.

Our attention will be restricted to FD and OD models, because we are interpreting conditional dependencies as demonstrated in Example 2.4, and under this interpretation the difference between conditional and unconditional models is unimportant.

3.1. Term semantics for FD models. In the preceding section we have shown that partial solutions of computational problems can be interpreted by acyclic and nonrecursive program schemes. This means that the function described by a partial solution is a superposition of functions corresponding to the dependencies of the model. Therefore, we obtain a description of solutions in the form of terms of finite functional types.

DEFINITION 3.1. Any variable A is a *type*. If α and β are finite sequences of types, then the expression $(\alpha \to \beta)$ is also a type.

The restrictions on the form of dependencies narrow the class of types required for our purposes to the functional types of the form $(W \to A)$ and operator types of the form $((W_1 \to A_1)\ldots(W_m \to A_m) \to (W \to A))$, where W_i, W are collections of variables and A_i, A are variables.

In the computational model Σ each functional type $(W \to A)$ is associated with the set $T^{(W \to A)}(\Sigma)$ of the terms of this type.

DEFINITION 3.2. Consider an FD model Σ. Then:
1. Every variable A in this model is a term of type $(A \to A)$, i.e., $A \in T^{(A \to A)}(\Sigma)$;
2. If $f:(A_1\ldots A_r \to A) \in \Sigma$ and $t_1 \in T^{(W_1 \to A_1)}(\Sigma), \ldots, t_r \in T^{(W_r \to A_r)}(\Sigma)$, then $f(t_1,\ldots,t_r)$ is a term of type $(\bigcup_1^r W_i \to A)$, i.e.,

$$f(t_1,\ldots,t_r) \in T^{(\bigcup_1^r W_i \to A)}(\Sigma).$$

Any expression that is a term according to Definition 3.2 will be called a *functional term*.

Under interpretation I of the symbols of the model Σ, each variable A is associated with its nonempty range D_I^A, the collection of variables $W = A_1\ldots A_r$ corresponds to the set of r-tuples $D_I^W = D_I^{A_1} \times \ldots \times D_I^{A_r}$, and the name f of the functional dependency $(W \to Y)$ to the function $f_1: D_1^W \to D_1^Y$. Therefore, under interpretation I, the functional term t of type $(W \to A)$ corresponds to the function $(\lambda W.t_1): D_1^W \to D_1^A$.

DEFINITION 3.3. Let $\Pi = (\Sigma; X, Z)$ be a computational problem and $A \in Z$. Functional terms in the set

$$\Phi^A(\Pi) = \bigcup_{W \subseteq X} T^{(W \to A)}(\Sigma)$$

will be called *partial A-solutions* of the problem Π. If a partial A-solution exists for each $A \in Z$, then the problem Π is (partially) solvable. If $Z = A_1\ldots A_m$, then an m-tuple of the terms $(t_1,\ldots,t_m)$ in which each term t_i is a partial A_i-solution of the problem Π is said to be a partial solution of this problem. The set of all partial solutions of the problem Π will be denoted by $\Phi(\Pi)$.

EXAMPLE 3.1. $\alpha_2(\beta_2(b,A), \sigma_2(A,b,c) \in T^{(Abc \to c)}(TRIANGLE_2)$, and the terms

$$\sigma_2(A,b,c) \in \Phi(\Pi)$$

and

$$\sigma_2(A, \alpha_1(\sigma_2(A,b,c), \beta_1(c,A)), \alpha_2(\sigma_2(A,b,c), \beta_2(b,A))) \in \Phi(\Pi)$$

are partial S-solutions of the problem $\Pi = (TRIANGLE_2; Abc, S)$ in the model $TRIANGLE_2$.

3.2. Algorithmic problems for FD models. We are now able to formulate algorithmic problems related to the search for partial solutions of computational problems in FD models.

Partial analysis problem: Given a computational problem Π, determine whether it has partial solutions, i.e., whether $\Phi(\Pi) \neq \emptyset$.

Partial synthesis problem: Given a computational problem Π, construct a partial solution $\bar{t} \in \Phi$ if the partial analysis problem is solved positively.

Equivalence problem: Given computational problems $\Pi_1 = (\Sigma_1; X, Z)$ and $\Pi_2 = (\Sigma_2; X, Z)$, determine whether their sets of partial solutions coincide, i.e., whether $\Phi(\Pi_1) = \Phi(\Pi_2)$.

The statement of the fourth problem requires additional discussion. Consider the problem $\Pi_1 = (TRIANGLE_2; abc, S)$ in the model $TRIANGLE_2$. One can check that problem Π_1 is equivalent to a problem in the smaller model $\Pi_2 = (\{\pi, \sigma_4\}; abc, S)$, i.e., the FD $\sigma_1, \sigma_2, \sigma_3, \alpha_1, \alpha_2, \beta_1, \beta_2$ are not necessary in solving the problem Π_1. This leads to the following notion.

DEFINITION 3.4. The computational problem $\Pi = (\Sigma; X, Z)$ is the *smallest* among the problems of the form $\Pi' = (\Sigma'; X, Z)$ equivalent to Π, if for any such problem Π' we have $\Sigma \subseteq \Sigma'$.

Clearly, for every computational problem there exists a unique smallest equivalent problem

Minimization problem. Given a computational problem $\Pi = (\Sigma; X, Z)$, construct a smallest problem of the form $\Pi' = (\Sigma'; X, Z)$ equivalent to Π.

In this brief survey, it is not our goal to describe efficient algorithms solving the problems stated above. Instead, we shall describe the methods of solving these problems on which the algorithms are based.

The most famous methods for solving the analysis and synthesis problems are those of the forward and backward chainings.

Forward chaining. In the computational problem $\Pi = (\Sigma; X, Z)$ each variable A is related to the number $r_x(A)$, the rank (of attainability from X) of the variable A in the problem Π defined as follows:

$$r_x(A) = 0 \quad \text{if } A \in X,$$

and

$$r_x(A) = \min_{f:(W \to A) \in \Sigma} \{1 + \max_{B \in W}\{r(B)\}, \omega\} \quad \text{if } A \notin X$$

(it is assumed that $\omega > i$ for any integer i, and $\omega + 1 = \omega$). One can show that r_x is a function defined everywhere on the set of values Σ. For example, it is not hard to check that in the computational model $TRIANGLE_2$ we have $r_{abc}(S) = 2$, $r_{Abc}(S) = 1$, and $r_{ABx}(S) = \omega$.

The following simple proposition holds.

PROPOSITION 3.1. (1) *The computational problem* $\Pi = (\Sigma; X, Z)$ *has a partial solution if and only if* $r_x(A) < \omega$ *for every variable* $A \in Z$.

(2) *If the rank of the variable A is finite (i.e., $r_x(A) < \omega$), then it does not exceed the total number k of variables in the model.*

This proposition provides the foundation for the forward chaining method. The method consists in consecutively constructing the sets $V^{(0)}, V^{(1)}, \ldots, V^{(k)}$ of all variables of rank 0, rank 1, etc., and verifying at each step i whether all goal variables are contained in the set $\bigcup_1^i V^{(j)}$. If at some step $i_0 \leq k$ this condition is satisfied, then the partial analysis problem is solvable; otherwise it is unsolvable. For example, when applied to the problem $(TRIANGLE_2; AbS, h_b)$, the forward chaining method produces the sets $V^{(0)} = \{A, b, S\}, V^{(1)} = \{h_c\}, V^{(2)} = \{c\}, V^{(3)} = \{h_b\}$, which indicates that the problem is solvable.

There are quite obvious algorithms solving the problems of analysis and synthesis via the forward chaining method in the time $O(|\Pi| \cdot p)$, where p is the number of variables in the computational model. In 1981 this author presented the algorithms of analysis and synthesis using the forward chaining method with linear running time $O(\Pi)$ on RAM with constant step weight. These algorithms first rewrite the computational problem (in linear time) in the form of a special kind of list structure, and then (again in linear time) solve the problem of analysis and synthesis on the list structure. The algorithms are published in [7,8]. The scheme of the analysis algorithm is given in the appendix.

Backward chaining. While the forward chaining algorithms look for a solution of computational problems starting from initial values and moving in the direction of goal values, the backward chaining algorithms operate in the opposite direction.

The backward chaining algorithms can be conveniently described in terms of the following model game on computational problems. Let $\Pi = (\Sigma; X, Z)$ be a computational problem. In the course of the game the variables of the model Σ are "covered with coins". For each collection of variables W the symbol W^h means that all the variables in W are covered by "heads", W^t means that all of them are covered by "tails", and W means that none of the variables is covered by a coin. A move in the game is an application of one of the following rules (the rules are formulated by stating the conditions of applicability first, and then the operation that can be performed under these conditions).

1. $\dfrac{A,\ A \in Z}{A^h}$ (the goal variable can be covered with "heads").

2. $\dfrac{A,\ B^h,\ f\colon (W_1 A W_2 \to B) \in \Sigma}{A^h}$ (if the variable at the conclusion of the FD is covered with "heads", then an uncovered variable in the premise can be covered with "heads").

3. $\dfrac{A^h, A \in X}{A^t}$ (if the initial variable is covered by "heads", the coin can be reversed).

4. $\dfrac{f\colon (W \to A) \in \Sigma, W^t}{A^t}$ (if all the variables in the premise of the FD are covered by "tails", then the variable in the conclusion can also be covered by "tails".)

Gain: Z^t.

An example of a winning strategy in the game on the problem $(TRIANGLE_2; abc, S)$ can be provided by the following sequence of steps (the rule numbers are indicated in parenthesis): $S^h(1), a^h(2), b^h(2), c^h(2), p^h(2), a^t(3), c^t(3), p^t(4), s^t(4)$.

The partial analysis problem is related to this game via the following proposition.

PROPOSITION 3.2. *A winning strategy in the game on the problem Π exists if and only if $\Phi \neq \emptyset$.*

A backward chaining algorithm is an algorithm of deterministic search for a winning strategy in the above game on a computational problem. The paper [8] presents a backward chaining algorithm solving the partial analysis problem in linear time $O(\Pi)$. This algorithm is easily upgraded to a synthesis algorithm which also has a linear run time.

We now turn to the problems of equivalence and minimization for computational problems. In order to solve these problems in polynomial time the following notion is useful.

DEFINITION 3.5. Let $\Pi = (\Sigma; X, Z)$ be a computational problem. An FD f is said to be *inseparable* from Π if there exists a sequence of FD $f_1, f_2, \ldots, f_n$ in Σ such that for each $f_i : (W_i \to A_i)$ all the values in W_i have finite ranks and $f_1 = f$, $A_n \in Z$, and $A_i \in W_{i+1}$ for all $1 \leq i \leq n$. The set of all FD inseparable from Π is denoted by $\text{insep}(\Pi)$.

For example, in the computational problems $\Pi_1 = (TRIANGLE_2; Abc, S)$, $\Pi_2 = (TRIANGLE_2; abc, S)$, and $\Pi_3 = (TRIANGLE_2; ABC, S)$ we have $\text{insep}(\Pi_1) = \{\sigma_2, \alpha_1, \alpha_2, \beta_1, \beta_2\}$, $\text{insep}(\Pi_2) = \{\sigma_4, \pi\}$, and $\text{insep}(\Pi_3) = \emptyset$.

THEOREM 3.3. *The computational problems $\Pi_1 = (\Sigma_1; X, Z)$ and $\Pi_2 = (\Sigma_2; X, Z)$ are equivalent if and only if $\text{insep}(\Pi_1) = \text{insep}(\Pi_2)$.*

THEOREM 3.4. *The computational problem $\Pi_1 = (\Sigma_2; X, Z)$ is the smallest problem equivalent to the problem $\Pi_2 = (\Sigma_2; X, Z)$ if and only if $\Sigma_1 = \text{insep}(\Pi_2)$.*

These two theorems yield algorithms solving the problems of equivalence and minimization in linear time $O(\Pi)$. Both algorithms are described in [8].

3.3. Term semantics of OD models. The first mathematical formulation of the problem of synthesis in computational problems with operator dependencies is contained in [9,10]. These papers use standard logical techniques. Dependencies of the form $\&_1^m (W_i \to Y_i) \to (W \to Y)$ and $(W \to Y)$ are understood as propositional formulas (the collection W of variables $A_1 \ldots A_k$ is interpreted as the conjunction $A_1 \& \ldots \& A_k$), the dependencies are understood as applied axioms, and a solution of the computational problem $\Pi = (\Sigma; X, Z)$ is understood as a proof of deducibility of $\Sigma \vdash (X \to Z)$ in an appropriate formal system for a positive fragment of propositional logic. As noted in [6,11], in such a formulation, the synthesis problem is a "search" problem. Namely, the results of [17,18] imply that it is *PSPACE*-complete.

Other formalizations of the synthesis problem in OD models are given in [5]. One of these formulations is due to the author and is based on the notion of the operator term; another is due to M. I. Kanovich and is based on the so-called lossless calculi. In terms of these calculi, it is convenient to note that the complexity of the synthesis problem in OD models is related to the degree of inseparability of subproblems. As in the example of computing an iterated integral, where one subproblem had to be solved in the framework of another, in the general case it may be necessary to order a set of subproblems in a sequence and solve them consecutively, using the conditions of all preceding subproblems in the solution of a given subproblem. If the solution depth, i.e., the minimal length of such sequences of problems, is not bounded by a single constant, then solving computational problems is a "search" problem; otherwise it is a problem of polynomial complexity. As was shown by Kanovich, the synthesis problem solvable with depth h has complexity $O(|\Sigma|^{h+2})$. For the case $h = 0$ the corresponding quadratic algorithm was first described by this author in 1983, and later published in [5].

In this survey we describe the term semantics of OD models for the separable subproblems mode. In defining term solutions in this mode for a computational problem in an OD model we have two new features as compared with the case of FD models. First, term solutions are formed not only by the superposition of functions but also by applying operators of appropriate type to functions. Second, one has to distinguish two modes of solving computational problems: with common memory and without common memory (see Example 2.2). The sets of term solutions of a given type for the same problem but different modes may turn out to be different.

DEFINITION 3.6. Consider a computational problem $\Pi = (\Sigma; X, Z)$ in the OD model Σ.

(1) Each variable A of the model Σ is a term of the type $(A \to A)$, i.e., $A \in T^{(A \to A)}(\Pi)$.

(2) (Functional superposition rule). If $f: (A_1 \ldots A_r \to A) \in \Sigma$ and $t_1 \in T^{(W_1 \to A_1)}(\Pi), \ldots, t_r \in T^{(W_r \to A_r)}(\Pi)$, then $f(t_1, \ldots, t_r) \in T^{(\bigcup_1^r W_i \to A)}(\Pi)$ (thus the first two conditions repeat Definition 3.2).

(3) (Operator superposition rule). If $F: \alpha_1 \ \& \ \ldots \ \& \ \alpha_r \to \alpha$ is an OD of the model Σ, and $t_1 \in T^{\alpha_1}(\Pi), \ldots, t_r \in \Sigma^{\alpha_r}(\Pi)$, then $F[t_1, \ldots, t_r] \in T^{\alpha}(\Pi)$.

(4) (Common memory rule). For any collection of initial values $Y \subseteq X$, any variable A, and any collection of variables W_1, W_2 we have

$$T^{(W_1 Y W_2 \to A)}(\Pi) = T^{(W_1 Y W_2 \to A)}(\Pi);$$

in other words, condition (4) defines an equivalence relation on the set of terms: the type of term does not depend on the insertion or deletion of initial variables in its premise. Denote by $RT^{\alpha}(\Pi)$ the set of terms of type α defined by (1)—(3). If $t \in RT^{\alpha}(\Pi)$, we say that term t has the type α in the narrow sense.

EXAMPLE 3.2. Consider the computational model Σ_0 with the dependencies

$$f: ad \to c, \qquad g: by \to d, \qquad F: (y \to c) \to (a \to c),$$

and the problem $\Pi_0 = (\Sigma_0; ab, c)$ in this model. It is not hard to see that

$$RT^{(ab \to c)}(\Pi_0) = \emptyset.$$

We have $T^{(ab \to c)}(\Pi_0) = \{F[f(a, g(b, y))]\}$. Indeed,

$$y \in T^{(y \to y)}(\Pi_0), \qquad b \in T^{(b \to b)}(\Pi_0)$$

and according to the functional superposition rule $g(b, y) \in T^{(by \to d)}(\Pi_0)$. Furthermore, $a \in T^{(a \to a)}(\Pi_0)$ and according to the same rule $f(a, g(b, y)) \in T^{(aby \to c)}(\Pi_0)$. herefore, it follows from the common memory rule that $f(a, g(b, y)) \in T^{(y \to c)}(\Pi_0)$.

Hence, by the operator superposition rule, we have $F[f(a, g(b, y))] \in T^{(a \to c)}(\Pi_0)$, and again by the common memory rule we have

$$F[f(a, g(b, y))] \in T^{(ab \to c)}(\Pi_0).$$

DEFINITION 3.6. Consider the computational problem Π in the OD model Σ. A term $t \in RT^{(W \to A)}(\Pi)$ is said to be a *partial A-solution* for the problem Π without common memory, or in the narrow sense (with separable subproblems), if $W \subseteq X$ and $A \in Z$. An r-tuple $t_1, \ldots, t_r$ of terms is said to be a *partial solution* of the problem Π without memory, or in the narrow sense (with separable subproblems), if t_i is a partial A_i-solution of Π for each $1 \leq i \leq r$ and $Z = A_1 \ldots A_r$. The set of all partial solutions of Π in the narrow sense is denoted by $RT(\Pi)$. The problem Π is (partially) solvable without common memory or in the narrow sense (with separable variables) if $RT(\Pi) \neq \emptyset$. A term $t \in T^{(X \to A)}$ is said to be a partial A-solution of the problem Π (with separable variables) if $A \in Z$. An r-tuple $(t_1, \ldots, t_r)$, where t_r is a partial A_i solution of Π for each $1 \leq i \leq r$, is a partial solution of Π if $Z = A_1 \ldots A_r$. Denote the set of all partial solutions of Π by $T(\Pi)$. The problem Π is (partially) solvable with common memory (with separable subproblems) if $T(\Pi) \neq \emptyset$.

Example 3.2 shows that the problem Π in it is partially solvable with common memory, but partially unsolvable without common memory. It is not hard to check that the problem $(ADVANCE; N\ RN\ S)$ is also partially unsolvable without common memory, while the problem $(ADVANCE; RN\ S)$ is partially solvable both with and without common memory (here, as we have agreed, the dependency $f_1: N\ RN \to P\ LE\ SX\ A\ M\ NC\ MS$ should be understood as an abbreviated notation for $f_1^P: N\ RN \to P$, $f_1^{LE}: N\ RN \to LE$, ..., etc.

Now we can formulate algorithmic problems for OD models similar to those for FD models. Here are the problems of analysis and synthesis for the mode with separable subproblems.

Partial analysis problem (partial analysis in the narrow sense): given a computational problem Π in an OD model Σ, determine whether it has partial solutions (respectively, partial solutions in the narrow sense), i.e., $T(\Pi) \neq \emptyset$ (respectively, $RT(\Pi) \neq \emptyset$).

Partial synthesis problem (partial synthesis in the narrow sense): given a computational problem Π in an OD model Σ, find a partial solution $\bar{t} \in T(\Pi)$ (respectively, a partial solution in the narrow sense $\bar{t} \in RT(\Pi)$) if the partial analysis problem (respectively, partial analysis in the narrow sense) is solved for Π positively.

The author has presented algorithms solving the analysis and synthesis problems (in particular, those in the narrow sense) in the class of OD models in the mode of separable subproblems with capacity and running time $O(S(\Pi))$, where S is the number of subproblems in the computational problem Π. These algorithms are published in [**5**].

For example, when applied to the iterated integral problem of Example 2.5, the partial analysis algorithm determines that the problem is unsolvable in the mode with separable subproblems, even with common memory.

The paper [5] contains equivalent formulations (due to M. I. Kanovich) of the partial analysis and synthesis problems in OD models, as well as efficient algorithms solving these problems in terms of lossless calculi.

4. Search for general solutions

As shown by examples in the second section, the necessity of searching for general solutions arises mainly for computational problems with conditional dependencies, especially in those cases where recursion is allowed. In this section, we consider precise formulations of the problems related to the search for general solutions in the class of CFD models.

A general solution of the computational problem in a CFD model arises naturally if we consider a CFD model as a recursive scheme. With this in mind, it is useful to modify the syntax for CFD models. We shall describe a CFD model in the form of a system of $E(\Sigma)$ formal equations of the form

$$(1) \qquad A = e_1 \square \ldots \square e_m \square Q_1 \to e_{m+1} \square \ldots \square Q_n \to e_{m+n},$$

where the e_i are either variables or expressions of the form $f(A_1, \ldots, A_r)$. An unconditional term of the form B or $f(A_1, \ldots, A_r)$ is present in equation (1) if and only if there is a unnamed FD in Σ of the form $(B \to A)$ or, respectively, a named FD of the form $f: A_1 \ldots A_r \to A$. A conditional term of the form $Q \to B$ or $Q \to f(A_1, \ldots, A_r)$ is present in equation (1) if and only if Σ includes an unnamed CFD of the form $Q \to (B \to A)$, or, respectively, a named FD of the form $f: Q \to (A_1 \ldots A_r \to A)$. In particular, this implies that for any A in the scheme $E(\Sigma)$ there is at most one equation defining A.

For example, if in the computational model *SORT* from Example 2.5 the dependencies $e_1: R \to S$ and $e_2: M \to R$ are respectively replaced by $R \to S$ and $M \to R$, the resulting model is described by the system $E(SORT)$ having the form

$$S = \mathrm{ins}(S, F) \square R, \qquad R = \mathrm{cdr}(R) \square M, \qquad F = \mathrm{car}_1(R) \square \mathrm{car}_2(M).$$

In order to simplify the notation, we shall consider computational problems with a single goal variable and shall write a problem in the form $\Pi = (E(\Sigma); X, A)$, where the model Σ is described by the scheme $E(\Sigma)$.

Having finished with syntactic conventions, we now proceed to the semantics of the schemes describing the computational models. Under a given interpretation of the symbols in the computational problem $(E(\Sigma); X, A)$ the semantics of the scheme $E(\Sigma)$ will be described as a way of calculating the values of the variable A for the given values of variables from X. This approach is absolutely natural: in order to compute the value of a variable B one can proceed as follows. In the equation

$$B = Q_1 \to e_1 \square \ldots \square Q_m \to e_m,$$

that defines B, find a term $Q_i \to e_i$ for which Q_i is true and set $B = e_i$, which reduces the problem of computing B to the computation of e_i. However, as we shall soon see, the class of such computations essentially depends on how the conditions Q_i are interpreted. We shall consider two different types of interpretation. The first of them, which would appear to be more natural, is that the conditions depend only on the variables in the model, or, in other words, they are relations in the sense of Definition 2.1. Unfortunately, such a narrow understanding of interpretation turns algorithmic problems related to synthesis into search problems. Another type of interpretation is that conditions are interpreted on a common carrier set Γ (we shall call it the set of states), which, generally speaking, is not necessarily related to the ranges of the variables. The first type of interpretation allows for no recursion in the computational model, and, therefore, such models are described by branching programs. Under the second type of interpretation recursion is possible, and we shall now describe a natural strategy for computing solutions recursively. The precise formulations are as follows.

4.1. General solutions of problems in branching models.

DEFINITION 4.1. A computational model $E(\Sigma)$ is said to be *branching* if, first, all conditions in the equations are of the form $R(W)$, where W is a nonempty set of variables; second, in each equation the names of the conditions are pairwise distinct; and, third, each equation has at most one unconditional term.

The paper [12] uses another term for branching models, which is, in our opinion, less suitable, viz., "acyclic models".

EXAMPLE 4.1. Consider the model $E(ADVANCE_3)$ from Example 2.4, which is close to $ADVANCE_2$. The dependencies in the model are of the form:

$$P = f_1^P(N, RN), \quad LE = f_1^{LE}(N, RN), \quad SX = f_1^{SX}(N, RN),$$
$$A = f_1^A(N, RN), \quad M = f_1^M(N, RN), \quad NC = f_1^{NC}(N, RN),$$
$$MS = f_1^{MS}(N, RN), \quad SA = f_2(P, LE), \quad IT = f_3(SA),$$
$$\% = C_1(SX, A, M, NC, MS) \rightarrow f_4(S) \square C_2(SX, A, M, NC, MS) \rightarrow g_4(S)$$
$$\square C_3(SX, A, M, NC, MS) \rightarrow h_4(S),$$
$$UST = f_6(S, \%), \quad S = f_5(SA, IT, UST).$$

Evidently, this model is a branching one.

The interpretation I of the branching model $E(\Sigma)$ defines for each variable B its nonempty range D_I^B, associates with the name of a dependency of the form $f: (\rightarrow B)$ or $f: Q \rightarrow B$, where Q is a condition, an object $f_I \in D_I^B$, and with the name of a dependency of the form $g: (W \rightarrow B)$ or $g: Q \rightarrow (W \rightarrow B)$, where $W \neq \emptyset$, a mapping $g_I: D_I^W \rightarrow D_I^B$, and with each variable B and relation of the form $R(W)$ a predicate R_{IB} on the set D_I^W.

DEFINITION 4.2. An interpretation I is said to be *deterministic* for a branching model $E(\Sigma)$ if for any of its equations $B = \ldots R_1(W_1) \rightarrow e_1 \square \ldots \square R_2(W_2) \rightarrow e_2 \ldots$ and for any relations $R_1(W_1)$ and $R_2(W_2)$ appearing therein the predicates $(R_1)_{IB}$ and $(R_2)_{IB}$ are incompatible (i.e., $(R_1)_{IB}$ & $(R_2)_{IB} \equiv 0$).

For example, it is not hard to see that the standard interpretation described in Example 2.4 is deterministic for the branching model $E(ADVANCE_3)$.

If in the branching model Σ there is a computational problem Π and a deterministic interpretation of Σ, then Π is implicitly related to a function-solution of the system of equations $E(\Sigma)$ which is computed in the following way.

DEFINITION 4.3. Let $E(\Sigma)$ be a branching model, $\Pi = (E(\Sigma); X, A)$ a computational problem in $E(\Sigma)$, I a deterministic interpretation for $E(\Sigma)$, and $\bar{d} \in D_I^X$

an n-tuple of initial values. For each variable B of the model $E(\Sigma)$, define the value $\Pi_I^B(d)$ as follows.

For all $B \in X$ set $\Pi_I^B(\bar{d}) = (\bar{d})_B$ (i.e., the B-component of the n-tuple $\bar{d}$; for example, if $X = BC$ and $\bar{d} = (b,c)$, then $(\bar{d})_B = b$). Denote by e_I the value, under the interpretation I, of the expression e in the equation term, i.e., the value $f_I(\Pi_I^{B_1}(\bar{d}), \ldots, \Pi_I^{B_k}(\bar{d}))$ if $e \stackrel{\circ}{=} f(B_1, \ldots, B_k)$, the value f_1 if $e \stackrel{\circ}{=} f$, and $\Pi_I^C(\bar{d})$ if $e \stackrel{\circ}{=} C$.

If $B \notin X$, the equation defining B is of the form $B = \ldots \Box R(A_1, \ldots, A_r) \to e \Box \ldots$, and the relation $R(A_1, \ldots, A_r)$ is true, i.e.,

$$R_{IB}(\Pi_I^{A_1}(\bar{d}), \ldots, \Pi_I^{A_r}(\bar{d})) = 1,$$

then $\Pi_I^B(\bar{d}) = e_I$. If $B \notin X$, the equation defining B is of the form $B = e \Box \ldots$ and includes no true relations, then $\Pi_I^B(\bar{d}) = e_I$. In the remaining cases $\Pi_I^B(\bar{d})$ is not defined.

Set $\Pi_I(\bar{d}) = \Pi_I^B(\bar{d})$ and call the partial function $\lambda x.\Pi_I$ a function-solution to the problem Π under the interpretation I.

EXAMPLE 4.2. Consider the problem $\Pi_a = (E(ADVANCE_3); N\ RN, S)$ in the model $E(ADVANCE_3)$ of the preceding example. Let us compute the value $\Pi_a(\text{Ivanova P. S.}, 007)$ in the standard interpretation.

$$\Pi_a(\text{Ivanova P. S.}, 007) = \Pi_a^S(\text{Ivanova P. S.}, 007)$$
$$= f_5(\Pi_a^{SA}(\text{Ivanova P. S.}, 007), \Pi_a^{IT}(\text{Ivanova P. S.}, 007), \Pi_a^{UST}(\text{Ivanova P. S.}, 007)).$$

Suppose that the first ten equations in our interpretation of $E(ADVANCE_3)$ yield the following values.

$\Pi_a^P(\text{Ivanova P. S.}, 007) = $ "engineer",

$\Pi_a^{LE}(\text{Ivanova P. S.}, 007) = 1.5$,

$\Pi_a^{SX}(\text{Ivanova P. S.}, 007) = $ "female",

$\Pi_a^A(\text{Ivanova P. S.}, 007) = 23$,

$\Pi_a^M(\text{Ivanova P. S.}, 007) = $ "married",

$\Pi_a^{NC}(\text{Ivanova P. S.}, 007) = 0$,

$\Pi_a^{MS}(\text{Ivanova P. S.}, 007) = $ "no",

$\Pi_a^{SA}(\text{Ivanova P. S.}, 007) = f_2(\Pi_a^P(\text{Ivanova P. S.}, 007), \Pi_a^{LE}(\text{Ivanova P. S.}, 007))$
$\qquad = f_2(\text{engineer}, 1.5) = 125$,

$\Pi_a^{IT}(\text{Ivanova P. S.}, 007) = f_3(\Pi_a^{SA}(\text{Ivanova P. S.}, 007))$
$\qquad = f_3(125) = 11.25$.

Then the tenth equation yields the relation

$$C_2\colon C_2(\text{"female"}, 23, \text{"married"}, 0, \text{"no"}) = 1.$$

Therefore

$$\Pi_a^{\%}(\text{Ivanova P. S.}, 007) = g_4(\Pi_a^{SA}(\text{Ivanova P. S.}, 007)) = g_4(125).$$

Suppose that under our interpretation $g_4(125) = 0.06$, i.e.,

$$\Pi_a^{\%}(\text{Ivanova P. S.}, 007) = 0.06.$$

Then

$$\Pi_a^{UST}(\text{Ivanova P. S.}, 007) = f_6(\Pi_a^{SA}(\text{Ivanova P. S.}, 007), \Pi_a^{\%}(\text{Ivanova P. S.}, 007))$$
$$= f_6(125, 0.06),$$

and if under our interpretation $f_6(SA, \%) = SA \cdot \%$, then

$$\Pi_a^{UST}(\text{Ivanova P. S.}, 007) = 125 \cdot 0.06 = 7.50.$$

Hence

$$\Pi_a(\text{Ivanova P. S.}, 007) = f_5(125, 11.25, 7.50).$$

If under our interpretation $f_5(SA, IT, UST) = SA \cdot 0.5 - (IT + UST)$, then, finally, we have $\Pi_a(\text{Ivanova P. S.}, 007) = 125 \cdot 0.5 - (11.25 + 7.50) = 43.75$.

The definition of a function-solution confronts us with the new situation: it turns a computational problem into a program in an algorithmic language. (In this language, the equation of the model is executed as a guarded Dijkstra's command, and the selection of the version $f(B_1, \ldots, B_r)$ calls the procedure that realizes the function f). This can create an impression that the new approach to computational problems makes the synthesis problem unnecessary. Of course, this is not the case. Recall that the synthesis problem also includes the problem of analysis, i.e., the problem of whether a solution exists. In our new situation the analysis problem becomes the problem of whether the function-solution is degenerate. However, even in this form the problem must be stated in more precise terms, since usually it is possible to find a "bad" interpretation I such that the function-solution Π_I is not defined anywhere. It is apparently reasonable to consider that the analysis problem is solved positively if under some interpretation I the function-solution is nondegenerate; otherwise the problem is considered to be solved negatively. While the negative solution to the analysis problem has the same meaning in this form as in the search for partial solutions, its affirmative solution has a weaker meaning. Indeed, a term solution, if any, defines a nondegenerate function-solution of a computational problem under any interpretation.

In practical applications this difference is not especially important, because in real life one always selects one standard application which makes it possible to make an adequate judgement on whether the problem is solvable.

So for the moment we shall consider the following formulation: *a computational problem* $\Pi(E(\Sigma); X, A)$ *has a general solution if under some interpretation I which is deterministic for the model $E(\Sigma)$, the function-solution Π_I is nondegenerate.* An evident drawback of this formulation is that it is not finitary. Let us present a more constructive one.

DEFINITION 4.4. Two computational models $\Pi_1 = (E(\Sigma_1); X, A)$ and $\Pi_2 = E(\Sigma_1); X, A)$ in branching models $E(\Sigma_1)$ and $E(\Sigma_2)$ are said to *generally equivalent* (and denoted $\Pi_1 =_{gb} \Pi_2$) if for any interpretation which is deterministic for both $E(\Sigma_1)$ and $E(\Sigma_2)$ the function-solutions Π_{1I} and Π_{2I} coincide.

DEFINITION 4.5. A computational problem $\Pi_0 = (E(\Sigma_0); X, A)$ in the branching model $E(\Sigma_0)$ is said to be the *smallest* of all problems which are generally equivalent to the problem $\Pi(E(\Sigma); X, A)$ in the branching model $E(\Sigma)$ if $\Pi_0 \equiv {}_{gb}\Pi$ and for any problem of the form $\Pi_1 = (E(\Sigma_1); X, A)$ in the branching model $E(\Sigma_1)$ such that $\Pi_1 \equiv {}_{gb}\Pi$ we have $\Sigma_0 \subseteq \Sigma_1$.

The desired finite restatement is given by the following theorem.

THEOREM 4.1. *Let $\Pi(E(\Sigma); X, A)$ be a computational problem in a branching model $E(\Sigma)$. A deterministic interpretation I for this model under which the function-solution Π_I is nondegenerate exists if and only if in the smallest problem $\Pi_0 = (E(\Sigma_0); X, A)$ generally equivalent to the problem Π the relation $\Sigma_0 \neq \emptyset$ holds.*

This theorem provides a motivation for the following main definition.

DEFINITION 4.6. Suppose that $\Pi_0 = (E(\Sigma_0); X, A)$ and $\Pi = (E(\Sigma); X, A)$ are problems in branching models, and let Π_0 be the smallest of all problems generally equivalent to Π. Then Π_0 will be called a *general solution* of the problem Π if $\Sigma_0 \neq \emptyset$.

Evidently, if a general solution exists, it is unique.

We are now in a position to give final formulations for the general analysis and synthesis problems in branching models.

General analysis problem: given a problem $\Pi(E(\Sigma); X, A)$, determine whether it has a general solution.

General synthesis problem: given a problem $\Pi(E(\Sigma); X, A)$, produce its general solution if such a solution exists.

We should not be perplexed by the fact that in some cases a general solution to a computational problem Π will turn out to be the problem itself. This is the case, for example, with the problem Π_a from Example 4.2. The point is that in real applications Definition 4.6 should be somewhat modified, defining a general solution not as the smallest generally equivalent problem with a nondegenerate function-solution, but as a program (in some preselected programming language) computing this function. This way is taken in [12, 14]. Thus, in real situations, solving the general synthesis problem includes planning, optimization, and translation.

Although branching models, by their form, are recursive schemes, it can be shown that the recursion in them is not expressible. This was shown in [14] in two steps: first one describes a natural mapping of computations of function-solutions of the computational problem onto the set of the so-called computation structures (which are analogs of terms), and then one demonstrates that the set of computation structures for a given computational problem is finite. Here we have no possibility to consider this aspect in more detail.

Let us now turn to the question of complexity of the general analysis and synthesis problems in branching models. One can present nondeterministic polynomial algorithms solving the problems of general equivalence, analysis, and synthesis (see [12]). These algorithms are based on the fact that two problems Π_1 and Π_2 are generally equivalent if and only if the sets of computation structures associated with them coincide. Unfortunately, the same fact also implies that these problems are NP-hard [12].

THEOREM 4.2. *The problems of analysis, synthesis, and general equivalence for branching models are NP-hard.*

The negative results (the search nature of the main algorithmic problems and nonexpressibility of recursion) demonstrate that the scheme for computing function-solutions given in Definition 4.2 is ill-suited for practical purposes. Therefore, we shall now introduce new techniques into this scheme, which will increase its expressibility and make it possible to define recursive computations. At the same time, the formulations of the analysis and synthesis problems will be left unchanged. As a result, the classes of general equivalence will be bigger, which means that the synthesis becomes less detailed. This will be compensated by a substantial increase in the number of solvable problems and higher efficiency of the analysis and synthesis algorithms.

4.2 General solutions of problems in recursive models. Before we proceed to the description of a new recursive semantics for CFD models, let us demonstrate an example of the recursion on which the new semantics is built. It is natural to call this recursion attributive, since the process of recursive computation is organized as a derivation in the attribute grammar [15,16].

EXAMPLE 4.3. Consider again the model $E(SORT)$. Introduce into the terms of the equations of this model the names of conditions for selecting these terms as follows:

$$S = NU \to \text{ins}(S, F) \square U \to R,$$
$$R = NZ \to \text{cdr}(R) \square Z \to M,$$
$$F = NZ \to \text{car}(R) \square Z \to \text{car}(M).$$

In the resulting model $E(SORT_1)$, consider the previous computation problem $(E(SORT); M, S)$. As is the case in the derivations of attribute grammar, the computation process for a function-solution in our problem will involve attributes. The role of the synthesized attributes (i.e., the attributes passed in the direction from M to S) will be played by the computable values of variables. The role of inherited attributes (i.e., the attributes passed in the direction from S to M) will be played by so-called states. A state serves to evaluate the conditions, so any object suitable for this purpose can be selected for this role. In formal language, the selection of the set of states is defined by the selection of an interpretation for the computational model. In our problem states are pairs of positive integers (m,n). A pair (m,n) describes an intermediate state of the process of the bubble sorting of the sequence of numbers of length $m+n$, the sorting process goes from right to left, and n is the length of the part of the sequence being sorted. The condition symbols U, NU, Z, NZ are interpreted by the following predicates on the set of states:

$$U(m,n) \equiv n = 1, \qquad NU(m,n) \equiv n \neq 1,$$
$$Z(m,n) \equiv m = 0, \qquad NZ(m,n) \equiv m \neq 0.$$

With each term in the equation we associate functions transforming the attributes. If the term of the equation is a variable (R or M), then the transforming functions do not alter the states. If the term in the equation is an expression (i.e., is of the form $\text{ins}(S, F)$, $\text{cdr}(R)$, $\text{car}(R)$, or $\text{car}(M)$), then it is associated both with functions computing the values of variables (i.e., synthesized attributes), and with functions that transform states (i.e., inherited attributes). The functions computing the values of variables are the same functions that are defined by the standard interpretation of the names ins, cdr, and car described in Example 2.5.

Let us describe the functions that transform states.

The following two functions are associated with the dependency name ins:

$$\text{ins}^{S \to S}(m, n) = (m+1, n-1),$$
$$\text{ins}^{S \to F}(m, n) = (m, n).$$

The name of cdr is associated with the function

$$\text{cdr}^{R \to R}(m, n) = (m-1, n+1),$$

and the name of car is associated with the identity functions

$$\text{car}^{F \to R}(m, n) = \text{car}^{F \to M}(m, n) = (m, n).$$

For example, the computation of the value of the variable S in the state (m, n) proceeds as follows. As a result of evaluating the predicates U and NU in the state (m, n), we select one of the terms of the equation defining S, e.g., the term $\text{ins}(S, F)$ (when $NU(m, n) = 1$). Then (recursively) we compute the value of S in the state $\text{ins}^{S \to S}(m, n)$ and the value of F in the state $\text{ins}^{S \to F}(m, n)$. Finally, apply ins to the values of S and F thus obtained, and $\text{ins}(S, F)$ becomes the new value of S in the state (m, n)

Thus, the selection of the term ins under condition NU defines two crossing flows of attributes (see Figure 4.1): the flow of values from S and F to S, and the flow of states from S to S and F.

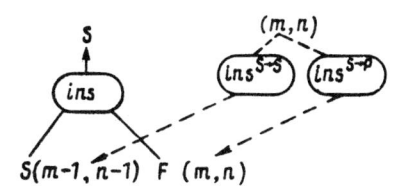

FIGURE 4.1

By way of example, we give the recursive computation tree for the value of S in the state $(0, 4)$, when the value for M is the array of numbers 5 1 3 2. The tree is shown in Figure 4.2, where for each variable the state for which this variable is computed is shown in parentheses with the computed value of the variable immediately below.

We now proceed to precise definitions.

DEFINITION 4.7. A computational model $E(\Sigma)$ is said to be *recursive* if (1) all conditions in its equations are represented only by their names; (2) in each equation the names of the conditions are pairwise distinct; and (3) each equation contains at most one unconditional term.

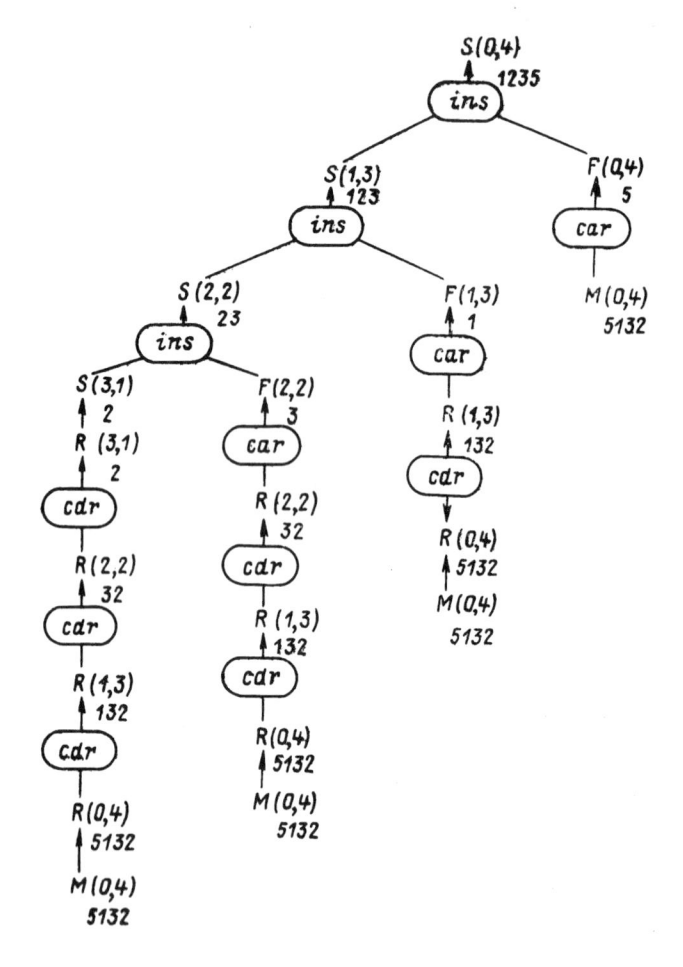

FIGURE 4.2

As seen from this definition, recursive models are syntactically different from branching ones only by the fact that they do not allow conditions in the form of relations. An example of a recursive model is the model $E(SORT_1)$ given above.

A recursive (R-)interpretation I for a recursive model $E(\Sigma)$ defines:

a nonempty set of states $\Gamma = \{q_1, q_2, \dots\}$,

for each variable B its nonempty range D_I^B and the mapping $h_b \colon \Gamma \times D_I^B \to D_I^B$ (the rule of passing a parameter when leaving the recursion),

for each condition name R the predicate R_I on the set of states Γ,

for each dependency name of the form $f \colon (\to B)$ or $f \colon R \to B$ the mapping $f_1 \colon \Gamma D_I^B$,

for each dependency name of the form $f \colon (B_1 \ldots B_r \to B)$ or $f \colon R \to (B_1 \ldots B_r \to B)$, where $r > 0$, the function $f_1 \colon D_I^{B_1 \ldots B_r} \to D_I^B$ and r mappings that transform states

$$f^{B \to B_1} \colon \Gamma \to \Gamma, \ldots f^{B \to B_1} \colon \Gamma \to \Gamma.$$

An example of an R-interpretation is the interpretation of the recursive model $E(SORT_1)$ partially described in Example 4.3. In order to complete the description of this interpretation, we have to add that the function h_M does not depend on states and is the identity mapping of D_I^M onto D_I^M, and that in all other cases the functions and the predicates can be defined arbitrarily.

DEFINITION 4.8. An R-interpretation I is said to be *deterministic* for a recursive model $E(\Sigma)$ if for each equation $B = \ldots \square R_1 \to e_1 \square \ldots \square R_2 \to e_2 \square \ldots$ of it and for any condition names R_1 and R_2 appearing in it, the predicates R_{1I} and R_{2I} are incompatible.

For example, the above R-interpretation of the model $E((SORT_1))$ is deterministic for this model.

A new recursive semantics for CFD models is contained in the following main definition.

DEFINITION 4.9. Let $E(\Sigma)$ be a recursive model, $\Pi(E(\Sigma); X, A)$ a computational problem in $E(\Sigma)$, I an R-interpretation that is deterministic for $E(\Sigma)$, $q \in \Gamma$ a state, and $d \in D_I^X$ an r-tuple of values of initial variables. For each variable B of the model $E(\Sigma)$ we define its value $\Pi_{Iq}^B(\bar{d})$ in the state q with the input $\bar{d}$ as follows.

Denote by $e_{1q}^B(\bar{d})$ the value of the expression e appearing in the term of the equation defining the variable B under the interpretation I in the state q with input $\bar{d}$, i.e., the value

$$f_I(\Pi_{1q_1}^{B_1}(\bar{d}), \ldots, \Pi_{Iq_r}^{B_r}(\bar{d})),$$

where $q_1 = f_I^{B \to B_1}(q), \ldots, q_r = f_I^{B \to B_r}(q)$ if $e \doteq f(B_1, \ldots, B_r)$;
the value $f_I(q)$ if $e \doteq f$, and
the value $\Pi_{Iq}^C(\bar{d})$ if $e \doteq C$.

For $B \in X$ set $\Pi_{Iq}^B(\bar{d}) = h_B(q, (\bar{d})_B)$ if either $E(\Sigma)$ contains no equation defining B, or the equation defining B is of the form $B = R_1 \to e_1 \square \ldots \square R_n \to e_n$ and all the conditions $(R_i)_I$ are false in the state q.

If the variable B is defined by an equation of the form

$$B = \ldots \square R \to e \square \ldots$$

and $R_I(q) = 1$ in it, or if B is defined by an equation of the form

$$B = e\square \ldots,$$

which either contains no conditional terms, or all the conditions are false in the state q, we set $\Pi_{1q}^B(\bar{d}) = e_{1q}^B(\bar{d})$.

In all other cases $\Pi_{Iq}^B(\bar{d})$ is not defined.

Set $\Pi_{Iq}(\bar{d}) = \Pi_{Iq}^A(\bar{d})$ and call the partial function $\lambda q X.\Pi_{Iq}$ a function-solution to the problem Π under the interpretation I.

EXAMPLE 4.4. Compute the value $\Pi_{(0,4)}^{(5132)}$ in the problem Π from Example 4.3 under the interpretation given therein. It is not hard to see that

$\Pi_{(0,4)}(5132)$
$= \mathrm{ins}(\mathrm{ins}(\mathrm{ins}(\Pi_{(3,1)}^R(5132), \Pi_{(2,2)}^F(5132)), \Pi_{(1,3)}^F(5132)), \Pi_{(0,4)}^F(5132))$
$= \mathrm{ins}(\mathrm{ins}(\mathrm{ins}(\mathrm{cdr}(\mathrm{cdr}(\mathrm{cdr}(5132))), \mathrm{car}(\mathrm{cdr}(\mathrm{cdr}(5132)))), \mathrm{car}(5132))$
$= \mathrm{ins}(\mathrm{ins}(\mathrm{ins}(2,3),1),5) = 1235.$

EXAMPLE 4.5. In this example we consider the problem of searching for a path in a labyrinth in the form of a solution to a computational problem in a recursive model. A labyrinth is a nonoriented graph whose vertices are positions and whose edges are transitions between positions. For each position the transitions from it to adjacent positions are linearly ordered. The problem of searching for a path in a labyrinth L is formulated as follows: given a pair of distinct positions e and 0, determine if there exists a path from e to 0 in L, and, if so, find such a path.

We start by describing the labyrinth L as a recursive model $E(\Sigma_L)$ in which every position b_i in L uniquely corresponds to the equation

$$b_i = Q \to \text{fl} \,\square R_{ij_1} \to \text{pass}_{ij_1}(b_{j_1})\square \ldots \square R_{ij_r} \to \text{pass}_{ij_r}(b_{j_r}),$$

where $b_{j_1}, \ldots, b_{j_r}$ are all the positions in L adjacent to the position b_i. The problem of searching for a path from position e to position 0 in L will be represented in the form of a computational problem $\Pi_L = (E(\Sigma_L); e, 0)$ in the following R-interpretation I_L.

(1) Let V be the set of all positions in the labyrinth L (i.e., the set of all variables in the model $E(\Sigma_L)$). Consider the alphabet $\tilde{V} = V \cup \{\bar{b} \mid b \in V\}$. Under the interpretation I_L, states are words in the alphabet V. A word-state q is always of the form wb, where b is the target position and w is the first segment of the path joining the initial and target positions. A symbol b_i appearing in the word w denotes the transition from the current position into position b_i. The symbol $\bar{b}_i$ denotes the return from position b_i. Denote by $[w]$ the irreducible word equal to the word $w \in \tilde{V}^*$ (the free group with generators V). Under our interpretation, for any state $q = wb$ the word $[w]$ represents the first segment of the path leading from the initial position in L without taking returns into account. If the symbol b is contained in the word w, this means that at some point the path passed through position b. A word w will be called a *dead-end word* if $[w] = w_1 b_i$ and the positions adjacent to b_i are already contained in the word w. A word w is said to be *leading to* b_j if $[w] = w_1 b_i$ and j is the least index among the positions that are adjacent to b_j but not contained in w.

(2) All the variables b_i have the same range $V^* \cup \{\text{fail}\}$. The parameter transition rule for the variable b_i is defined by the equality $h_{b_i}(wb_i) = [w]$, where wb_i is a state.

(3) For each i the predicates in the equation describing the position b_i are defined as follows:

$Q_i(wb) = 1$ if and only if $[w] = b_i$, $b_i \neq b$, and the word w is a dead-end one.

If j is the index of the position adjacent to b_i, then $R_{ij}(wb) = 1$ if and only if $[w] = w_1 b_i$ for some w_1, $b_i \neq b$, $Q_i(wb) = 0$, and the word w either leads to b_j, or is a dead-end one, but $[wb_i] = w_2 b_j$ for some w_2.

(4) $\text{fl}(q) = \text{fail}$.

(5) For each position b_i and each position b_j adjacent to it the function pass_{ij} is the identity and

$$\text{pass}_{ij}^{b_i \to b_j}(wb) = \begin{cases} wb_j b, & \text{if } b_j \text{ is not contained in } w, \\ w\bar{b}_i b, & \text{otherwise,} \end{cases}$$

It is not hard to check that the R-interpretation I_L is deterministic for the model $E(\Sigma_L)$. One can show that under this interpretation $e(\Pi_L)_{e0}(x) = \text{fail}$ for

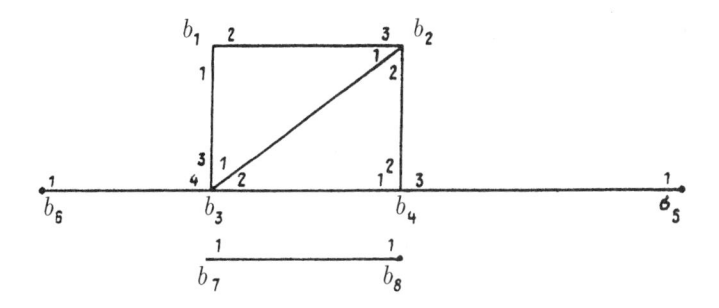

FIGURE 4.3

any x if there is no path in L leading from the initial position e to the target position 0, and $(\Pi_L)_{e0}(x)$ is a path from e to 0 if such paths exist.

For example, if the labyrinth L is of the form shown in Figure 4.3, then for the problem $\Pi_1 = (E(\Sigma_L); b_1, b_6)$ we have

$$(\Pi_1)_{b_1 b_6} a(\Lambda) = \text{pass}_{13}(\text{pass}_{32}(\text{pass}_{24}(\text{pass}_{45}(\text{pass}_{54}$$
$$\times (\text{pass}_{42}(\text{pass}_{23}(\text{pass}_{36}(h_{b_6}(b_1 b_3 b_2 b_4 b_5 \bar{b}_5 \bar{b}_4 \bar{b}_2 b_6 b_6) \ldots) = b_1 b_3 b_6,$$

and for the problem $\Pi_2 = (E(\Sigma_L); b_6, b_7)$ we have

$$(\Pi_2)_{b_6 b_7}(\Lambda) = \text{pass}_{63}(\text{pass}_{32}(\text{pass}_{24}(\text{pass}_{45}(\text{pass}_{54}$$
$$\times (\text{pass}_{42}(\text{pass}_{21}(\text{pass}_{12}(\text{pass}_{23}(\text{pass}_{36}(\text{fl}) \ldots) = \text{fail}.$$

DEFINITION 4.10. Computational problems $\Pi_1 = (E(\Sigma_1); X, A)$ and $\Pi_2 = (E(\Sigma_2); X, A)$ in recursive models $E(\Sigma_1)$ and $E(\Sigma_2)$ are said to be *generally equivalent* (denoted $\Pi_1 \equiv \Pi_2$) if for any R-interpretation deterministic for both $E(\Sigma_1)$ and $E(\Sigma_2)$ the function-solutions $\lambda q X.\Pi_1$ and $\lambda q X.\Pi_2$ coincide.

DEFINITION 4.11. A computational problem $\Pi_0 = (E(\Sigma_0); X, A)$ in the recursive model $E(\Sigma_0)$ is said to be the *smallest* of all problems generally equivalent to the problem $\Pi(E(\Sigma); X, A)$ in the recursive model $E(\Sigma)$ if $\Pi_0 \equiv \Pi$ and for any problem of the form $\Pi_1 = (E(\Sigma_1); X, A)$ in the recursive model $E(\Sigma_1)$ such that $\Pi_1 \equiv \Pi$ we have $\Sigma_0 \subseteq \Sigma_1$.

DEFINITION 4.12. Let $\Pi_0(E(\Sigma_0); X, A)$ and $\Pi = (E(\Sigma); X, A)$ be problems in recursive models, and Π_0 the smallest of all problems generally equivalent to Π. Then Π_0 is said to be a *general solution* to the problem Π if $\Sigma_0 \neq \emptyset$.

General analysis problem: given a problem in a recursive model, determine whether it has a general solution.

General synthesis problem: given a problem in a recursive model, present its general solution if such a solution exists.

EXAMPLE 4.6. Consider a more complicated version of the problem from Example 4.5. Suppose that positions in the labyrinth L are used to store integer variables, and the problem is to increase the input integer by the value of the sum of all variables positioned along some path leading from the input position to the output position (if such a path exists). All the parameters in the problem are subject to change: the input and output positions, the input integer, and the values of the variables at the positions in the labyrinth.

The difference between the new recursive interpretation and the one described in Example 4.5 is that the values of variables now are either integers or fail, and that the functions h_b and pass_{ij} are defined in a new way: $h_{bi}(q,n) = n + \text{`}b_i\text{'}$ (here q is a state, n the input value, and 'b_i' the value of the variable stored at position b_i);

$$\text{pass}_{ij}(x) = \begin{cases} x + \text{`}b_j\text{'}, & \text{if } x \neq \text{fail,} \\ \text{fail}, & \text{if } x = \text{fail}. \end{cases}$$

If the labyrinth L is of the form presented in Figure 4.3, then the general solution to the computational problem Π_1 of Example 4.5 is the problem $\Pi_{01} = (E(\Sigma_{01}); b_1, b_6)$ in the model $E(\Sigma_{01})$ describing the connected component in L consisting of the vertices $b_1, b_2, b_3, b_4, b_5, b_6$. It is not hard to check that

$$(\Pi_{01})_{b_1 b_6}(n) = (\Pi_1)_{b_1 b_6}(n) = n + \text{`}b_1\text{'} + \text{`}b_3\text{'} + \text{`}b_6\text{'}.$$

The complexity bounds for the problems of analysis, synthesis, and equivalence in the class of recursive models were obtained in [12] with the help of the following reduction.

With each recursive model $E(\Sigma)$ we associate an FD model Σ^n by deleting condition names from all conditional FD. This transforms every dependency π of the model Σ into the dependency π^n of the model Σ^n. For a problem $\Pi = (E(\Sigma); X, A)$ denote by Π^n the problem $(\Sigma^n; X, A)$ in the FD model Σ^n, and set

$$\text{insep}(\Pi) \rightleftharpoons \{\pi^n \in \Sigma \mid \pi^n \in \text{insep}(\Pi^n)\}.$$

THEOREM 4.3. *For any computational problems* $\Pi_1 = (E(\Sigma_1); X, A)$ *and* $\Pi_2 = (E(\Sigma_2); X, A)$ *in recursive models we have*

$$\Pi_1 \equiv \Pi_2 \iff \text{insep}(\Pi_1) = \text{insep}(\Pi_2).$$

THEOREM 4.4. *A problem* $\Pi'_0 = (E(\Sigma_0); X, A)$ *is the smallest of the problems generally equivalent to the problem* $\Pi = (E(\Sigma); X, A)$ *in a recursive model* $E(\Sigma)$ *if and only if*

$$E(\Sigma_0) = \text{insep}(\Pi).$$

These theorems have the following useful corollaries.

COROLLARY 4.5. *Two problems* Π_1 *and* Π_2 *in recursive models are generally equivalent if and only if so are the corresponding problems* Π_1^n *and* Π_2^n *in FD models.*

COROLLARY 4.6. *A problem* Π_0 *is the smallest of all problems generally equivalent to the problem* Π *in a recursive model if and only if the problem* Π_0^n *is the smallest of the problems equivalent to* Π^n.

This reduction demonstrates that the problems of general analysis and synthesis for recursive models are solvable in linear time $O(|\Pi|)$ with the help of the minimization algorithm from [8], while the problem of general equivalence for problems in recursive models is solvable in linear time $O(|\Pi_1| + |\Pi_2|)$ with the help of another algorithm from [8] recognizing the equivalence of models in FD models.

Furthermore, a solution of the general analysis problem for recursive models can be obtained with the help of the partial analysis algorithm for FD models, as the following statement shows.

THEOREM 4.7. *A problem Π in a recursive model has a general solution if and only if the problem Π^n in the corresponding FD model is partially solvable.*

We conclude this paper with a brief discussion of the formulations of the analysis and synthesis problems.

(1) Even in the case where the computational model does not change (which is quite possible in practical situations), the algorithms of analysis and synthesis can still solve nondegenerate problems because they can be applied to different computational problems in the model Σ. Going from one problem in a given model to another does not rebuild the structures on which these algorithms work; the collections of input and output values for a problem are parameters of algorithms that do not affect their complexity.

(2) From the viewpoint of realizability, the problems of searching for partial and general solutions in FD models and branching models respectively are, as a whole, similar: in both cases one has to provide program realizations for the algorithm synthesizing the control modules, as well as all library modules realizing individual functional dependencies in a given interpretation. A more creative task is, of course, the realization of search for general solutions to computational problems in recursive models, because, in addition, one has to select and provide program realizations for all state transforming functions. Here the selection of state transforming functions is a problem similar to finding the invariants of cycles in iterative programs.

(3) How should we view the search character of the algorithmic problems related to synthesis in the class of branching models, and linearity of the same problems in the class of recursive models? For example, should we think that the notion of a branching model is inherently unsuitable and inapplicable to practical problems? We do not think so. In the author's opinion, these results demonstrate that in order to make synthesis in branching models more effective it is sufficient (and desirable) to sacrifice the ability of the synthesizing algorithms to recognize ($\equiv_{gb}$) the nonequivalence of problems. This ability is of purely theoretical significance, and it is not easy (as was, for example, done in Theorem 4.2) to find a pair of computational problems that are generally equivalent in the class of recursive models, but not in the class of branching models. Therefore, in practical situations we can regard a branching model as a recursive one and apply effective algorithms of analysis and synthesis. The resulting solution will mostly coincide with the solution obtained by search algorithms for branching models.

5. Appendix. Schematic algorithm analyzing a computational problem by the forward chaining method

5.1. List structure describing a computational problem.

(a) *SOURCES* is the list of input values; *GOALS* is the list of output values. For each FD π *PREMISE*(π) is the list of values in its premise, and *CONCLUSION*(π) is the list of values in its conclusion. For each variable b *OCCURRENCES*(b) is the list of all occurrences of this variable in the lists *SOURCES*, *GOALS*, and the lists of the form *PREMISE*(π).

(b) It is possible to set up pointers within the above lists and between them in such a way that the operations

include(Element, LIST), which includes an element in the list,
exclude(Element, LIST), which (temporarily) excludes an element from the list,
restore(Element, LIST), which restores an excluded element,
$LIST = \emptyset$, which checks whether the list is empty,
$LIST_1 \cap LIST_2 \ne \emptyset$, which checks whether the intersection of two lists is empty,
are performed in time bounded by a (very small and depending on the language of realization) constant C_0 in any algorithmic language having pointers, provided that pointer transition, as well as the verification of whether pointers or variables coincide, corresponds to the same step.

(c) The list structure representing a computational problem Π is constructed from a linear representation of Π in $C_0 \cdot |\Pi|$ steps, and the same time is required to restore this structure after the algorithm halts.

(d) The list structure representing a problem Π is incremental in the sense that the addition to Π (or removal from Π) of a dependency π is performed in time $C_0 \cdot |\pi|$.

5.2. An analysis algorithm for a computational problem represented by a list structure.

```
for b ∈ SOURCES
do { exclude(b, SOURCES) ;
    if OCCURRENCES(b) ≠ ∅
    then { if OCCURRENCES(b) ∩ GOALS ≠ ∅
            then { exclude(b, GOALS) ;
                    if GOALS = ∅
                    then { success }
                } ;
            for c ∈ OCCURRENCES(b)
            do { exclude(b, PREMISE(π)) ;
                -- for π incident to this occurrence of b
                if PREMISE(π) = ∅
                then { for d ∈ CONCLUSION(π)
                        do { include(d, SOURCES) }
                    }
                }
        } ;
    if SOURCES = ∅
    then{ fail }
}
```

5.3. This algorithm terminates successfully if and only if the problem represented by the list structure is solvable.

5.4. The algorithm was repeatedly used in systems of automated program synthesis and mechanical theorem proving (see, for example, [20]).

References

1. E. H. Tyugu, *Solution of problems over computational models*, Zh. Vychisl. Mat. i Mat. Fiz. **10** (1970), 716–733; English transl in USSR Comput. Math. and Math. Phys. **11** (1970).
2. A. Heyting, *Intuitionism*, North Holland, Amsterdam, 1971.
3. A. N. Kolmogorov, *Zur Deutung der intuitionistischen Logik*, Math. Z. **35** (1932), 58–65.
4. S. C. Kleene, *Introduction to metamathematics*, Van Nostrand, New York, 1952.
5. A. N. Dikovsky and M. I. Kanovich, *Computational models with separable subproblems*, Izv. Akad. Nauk SSSR Tekhn. Kibernet. **1985**, no. 5, 36–59; English transl in Soviet J. Comput. Systems Sci. **24** (1986), no. 3, 14–34.
6. G. E. Mints and E. H. Tyugu, *Structural synthesis and nonclassic logics*, Third Conf. "Application of Methods of Mathematical Logic", Abstracts of Reports, Tallinn, 1983, pp. 52–60. (Russian)
7. A. N. Dikovsky, *Nets on coverings and analysis of functional dependencies*, Automata, Algorithms, Languages, Kalinin, 1982, pp. 69–85. (Russian)
8. _____, *Linear time solution of algorithmic problems related to the synthesis of acyclic programs*, Programmirovanie **1985**, no. 3, 38–49; English transl. in Programming and Computer Software **11** (1985).
9. G. E. Mints and E. H. Tyugu, *Completeness of the rules of structural synthesis of acyclic programs*, Dokl. Akad. Nauk SSSR **263** (1982), 291–294; English transl in Soviet Math. Dokl. **25** (1982).
10. _____, *Justification of the structural synthesis of programs*, Automatic Program Synthesis, Tallinn, 1983, pp. 5–40. (Russian)
11. B. B. Volozh, M. B. Matskin, G. E. Mints, and E. H. Tyugu, *The PRIZ system and propositional calculus*, Kibernetika (Kiev) **1982**, no. 6, 63–70; English transl. in Cybernetics **18** (1982).
12. A. N. Dikovsky, *Deterministic computational models*, Izv. Akad. Nauk SSSR Tekhn. Kibernet. **1982**, no. 6, 63–70; . English transl., Engrg. Cybernetics **22** (1984), no. 6, 121–136.
13. E. Dijkstra, *A discipline of programming*, Prentice-Hall, Englewood Cliffs, NJ, 1976.
14. A. N. Dikovsky, *Optimal algorithm for synthesizing optimal algorithms over computational models*, Optimization and Program Transformation (All-Union Seminar). Part 1, Novosibirsk, 1983, pp. 140–152. (Russian)
15. D. E. Knuth, *Semantics of context-free languages*, Math. Systems Theory **2** (1968), 127–145.
16. Ph. Lewis, D. Rosenkrantz, and R. Stirns, *Compiler design theory*, Addison-Wesley, Reading, MA, 1976.
17. R. E. Ladner, *The computational complexity of provability in systems of modal propositional logic*, SIAM J. Comput. **6** (1977), 467–480.
18. R. Statman, *Intuitionistic propositional logic is polynomial-space complete*, Theoret. Comput. Sci. **9** (1979), 67–72.
19. M. I. Kanovich, *Effective logical algorithms for the analysis and synthesis of dependencies*, Dokl. Akad. Nauk SSSR **285** (1985), 1301–1305; English transl in Soviet Math. Dokl. **32** (1985).
20. I. O. Babaev, S. S. Lavrov, G. A. Netsvetaeva, F. A. Novikova, and G. M. Shuvalov, *SPORA, a programming system with automatic program synthesis*, Third Conf. "Application of Methods of Mathematical Logic", Abstracts of Reports, Tallinn, 1983. (Russian)

Effective Calculi as a Technique for Search Reduction

M. I. Kanovich

ABSTRACT. A technique is proposed for search reduction and constructing effective algorithms in a number of problems connected with relational databases, program synthesis, deduction search, graph problems, etc. The technique is based on the use of appropriate logical calculi.

Constructing an effective (with polynomial run time) algorithm which solves problems of a given class requires *either reduction or elimination of search* in the problems of the class under consideration.

In a number of problems connected with relational databases, program synthesis, deduction search, graph problems, etc., one can reduce search and obtain efficient algorithms making use of appropriate logical calculi.

The main idea of this approach is as follows. One constructs a calculus S such that

1. a formula "ζ" expressing the problem ζ in the language of the calculus S is deducible in S if and only if problem ζ has a solution (the *completeness of the calculus S* with respect to the given class of problems);
2. in the deduction (counter-deduction) trees of the calculus S there are no dead-end branches, i.e., the deduction has *no loss of information*;
3. deduction sizes are not big: every deduction step requires a limited time (the calculus S is *effective*);
4. a solution to the problem ζ is easily constructed from the deduction of formula "ζ" in S.

Consider the algorithm ρ solving a problem ζ in the following way: first (e.g., by a reverse application of inference rules) one constructs a "deduction" tree ("reverse deduction" tree), and then it is checked whether the formulas located at the hanging points are axioms. The result gives the answer to the question of solvability for problem ζ, and the deduction serves as the basis for obtaining the solution ζ.

The *correctness of the algorithm* ρ follows from (1) and (2). The global correctness is reduced to the verification of local steps in the deduction, which makes it possible to exercise semantically reasonable control over intermediate results.

The *effectiveness of the algorithm* ρ is guaranteed by (3) and (4).

©1996 American Mathematical Society

The problem of search reduction in such lossless calculi is solved in two steps:
1. redundant inspections and returns are completely eliminated;
2. the number of required branches is restricted.

In this paper we demonstrate an application of this approach to two specific problems. The run times of the resulting algorithms are *linear* or *subquadratic*.

In Section 1 we consider the problem of recognizing the deducibility of the functional, implicit, multivalued, or complete hierarchical dependency D in the system of functional, implicit, multivalued, and complete hierarchical dependencies of Σ. This problem is of interest in the theory of relational databases, and in the theory of program synthesis allowing for nondeterministic and parallel computations. The main result consists in the construction of the algorithm ρ with *linear run time* which "builds" the dependency $X \to B$ out of the initial dependencies of different kind in Σ (Theorems 1.23 and 1.25; cf. [4]).

Note that in Section 1 we introduce a new type of dependency, viz. the *implicit dependency*, and that the idea of the $F - L - MVD$ calculus makes it possible to construct the *complete calculus* containing other types of dependencies: join dependencies and template dependencies. Unfortunately, it turns out to be impossible to guarantee (3) (which comes as no surprise since the original problem is NP-hard).

The algorithms α and ρ in Section 1 serve as components in many algorithms in the theory of relational databases. The results of this section make it possible to obtain *effective polynomial algorithms* of that kind with a clear semantic interpretation of their performance.

Section 2 considers the problem of schematic synthesis of minimal programs. The main result is as follows:
1. in Theorem 2.6 a (time) subquadratic synthesis algorithm is obtained for programs with minimal size;
2. in Corollary 2.4 and Theorem 2.5 it is proved that the synthesis problem for programs with minimal run time is NP-complete.

The proposed $F - LD$ calculus clearly yields a linear run time algorithm recognizing satisfiability and tautology in generalized Horn's propositional formulas.

A modification of this algorithm yields a *linear run time algorithm* recognizing satisfiability for 2-CNF (cf. [5]).

Appropriate lossless calculi make it possible to obtain effective algorithms even for search problems. In many cases the class of problems ζ has a stratification with respect to some parameter τ so that the algorithm solving ρ and based on a deduction search in the lossless calculus turns out to be "quasipolynomial": its run time when solving the problem ζ does not exceed $O(|\zeta|^{cr})$, where r is the value of the parameter τ for the problem ζ. Similar quasipolynomial algorithms based on lossless calculi are also obtained for the problem of satisfiability in propositional formulas, program synthesis in computational models with subproblems, some problems on graphs, etc.

1. Analysis of functional, implicit, and multivalued dependencies

In this section we use lossless calculi to obtain fast analysis algorithms for functional, implicit, and multivalued dependencies in relational databases.

We start with the necessary definitions.

Let $U = \{A_1, A_2, \ldots, A_k\}$ be a set of variables (attributes). All attributes are distinct.

The letters X, Y, Z, W, R, H (possibly with indices) will denote subsets of U. Denote by XY the union of X and Y. In place of $X \cup \{A\}$ we shall write XA. The set $U \setminus X$ will be denoted by X^c.

By a *relation scheme* we mean the expression $R(U)$ [1]. Here the set U is understood as a list of variables whose order can be ignored since the names of the variables are unique.

An example of the relation scheme is

(1) $$\text{Triangle } (a, b, c, \alpha, \beta, \gamma, S).$$

An *interpretation of the scheme* $R(U)$ is an arbitrary predicate (relation) $r(U)$; the domains of the variables $A_1, A_2, \ldots, A_k$ do not necessarily coincide. An interpretation $r(U)$ can be described by the table T_r whose columns are labeled by the names $A_1, A_2, \ldots, A_k$, while the rows are all the k-tuples $a = (a_1, a_2, \ldots, a_k)$ for which $r(a)$ is satisfied.

An example of the interpretation table for the scheme (1) is

(2)

a	b	c	α	β	γ	S
1	1	1	60	60	60	$\frac{\sqrt{3}}{4}$
2	2	2	60	60	60	$\sqrt{3}$
1	1	$\sqrt{2}$	45	45	90	$\frac{1}{2}$
2	$2\sqrt{2}$	2	45	90	45	2

This example shows that the attributes in the relation scheme (1) can be understood as "side 1", "side 2", "side 3", "angle 1", "angle 2", "angle 3", "area" related by the scheme "*Triangle*".

An expression $X \to Y$ is called a *functional dependency* [1].

A functional dependency $X \to Y$ is said to be *satisfied in the interpretation* $r(U)$ if there are no two rows in the table T_r whose components coincide with respect to all attributes from X but differ with respect to an attribute from Y [1].

The expression $\text{Circ}(X)$ will be called an *implicit* (or *circular*) *dependency*. An implicit dependency $\text{Circ}(X)$ is said to be *satisfied in* $r(U)$ if there are no two rows in the table T_r whose components differ with respect to an attribute from X but coincide with respect to all other attributes from X.

An expression $X \to\to U$ is called a *multivalued dependency* [1]. A multivalued dependency $X \to\to Y$ is said to be *satisfied in* $r(U)$ if the implication

$$(\exists Z r(U) \ \& \ \exists Z' r(U)) \supset r(U)$$

is identically true. Here $Z = Y \setminus X$, $Z' = Y^c \setminus X$ (cf. [1]).

For example, the following dependencies are satisfied in (2):

$$\text{Circ}(\alpha\beta\gamma), \quad \text{"the sum of angles theorem"},$$
$$ab\gamma \to c, \quad \text{"the sine theorem"},$$
$$abc \to \gamma, \quad \text{"the sine theorem"},$$
$$ab\gamma \to S,$$
$$abc \to S, \quad \text{"Heron's formula"},$$
$$\alpha\beta \to\to abcS.$$

A dependency D *logically follows* from the list of dependencies Σ if in any finite interpretation $r(U)$ in which all the dependencies in Σ are satisfied, the dependency D is also satisfied.

For example, "Heron's formula" $abc \to S$ logically follows from the list of dependencies $abc \to \gamma$, $ab\gamma \to S$.

A *formula* in the calculus we are now constructing is an expression of the form

$$\Sigma \Rightarrow D,$$

where D is a dependency and Σ is a list (finite set) of dependencies.

The *semantic conventions* are as follows. A formula $\Sigma \Rightarrow D$ is said to be true if D logically follows from Σ.

The $F - L - MVD$ calculus.

The axioms:

$$\Sigma \Rightarrow (XY \to Y),$$
$$\Sigma \Rightarrow (XY \to\to Y),$$
$$\Sigma \Rightarrow (XY \to\to Y^c).$$

The inference rules are stated pairwise using the symbol $\stackrel{*}{\to}$. The first element in the pair is obtained if we replace $\stackrel{*}{\to}$ with $\to$, and the second inference rule is obtained if we replace $\stackrel{*}{\to}$ with $\to\to$.

Inference rules:

Weakened FD:

$$\frac{\Sigma, (X \to Y) \Rightarrow (WA \stackrel{*}{\to} Z)}{\Sigma, (XA \to Y) \Rightarrow (WA \stackrel{*}{\to} Z)}.$$

Weakened LD:

$$\frac{\Sigma, \text{Circ}(X) \Rightarrow (WA \stackrel{*}{\to} Z)}{\Sigma, \text{Circ}(XA) \Rightarrow (WA \stackrel{*}{\to} Z)}.$$

Weakened MVD:

$$\frac{\Sigma, (X \to\to Y) \Rightarrow (WA \stackrel{*}{\to} Z)}{\Sigma, (XA \to\to Y) \Rightarrow (WA \stackrel{*}{\to} Z)}.$$

Introduction of FD:
$$\frac{\Sigma \Rightarrow (WY \stackrel{*}{\to} Z)}{\Sigma, (\to Y) \Rightarrow (W \stackrel{*}{\to} Z)}.$$

Introduction of LD:
$$\frac{\Sigma \Rightarrow (WA \stackrel{*}{\to} Z)}{\Sigma, \mathrm{Circ}(A) \Rightarrow (W \stackrel{*}{\to} Z)}.$$

Introduction of MVD:
$$\frac{\Sigma \Rightarrow (WY \stackrel{*}{\to} Z)\Sigma \Rightarrow (WY^c \stackrel{*}{\to} Z)}{\Sigma, (\to\to Y) \Rightarrow (W \stackrel{*}{\to} Z)}.$$

Here Σ is the set of dependencies, whose ordering we ignore; A is an attribute.

THEOREM 1.1. *The $F - L - MVD$ calculus is semantically correct.*

The proof is achieved by the direct verification of correctness for all inference rules.

THEOREM 1.2. *All inference rules in the $F-L-MVD$ calculus are reversible.*

The deduction search for a formula $\Sigma \Rightarrow D$ will be done as follows. Apply all the inference rules "in reverse" in an arbitrary order *so long as it is possible*. The resulting trees will be called reverse deduction trees for $\Sigma \Rightarrow D$. Naturally, different orders in the reverse application of inference rules result in different reverse deduction trees.

Suppose that the hanging vertices of the reverse deduction tree for $\Sigma \Rightarrow (X \stackrel{*}{\to} Z)$ are associated with the following formulas:

(3) $\qquad \Sigma_1(X_1 \stackrel{*}{\to} Z), \Sigma_2 \Rightarrow (X_2 \stackrel{*}{\to} Z), \ldots, \Sigma_p \Rightarrow (X_p \stackrel{*}{\to} Z).$

The fact that the inference rules are correct and reversible ensures that there are no dead-end branches and that there is no loss of information at any point of the tree.

COROLLARY 1.3. *A formula $\Sigma \Rightarrow (W \stackrel{*}{\to} Z)$ is true if and only if all the formulas in the list (3) are true.*

A formula $\Sigma \Rightarrow (W \stackrel{*}{\to} Z)$ is said to be *primitive* if for any dependency $(X \to Y)$, $(X \to\to Y)$, or $\mathrm{Circ}(X)$ from Σ we have $X \neq \emptyset$, $|X| \geq 2$ for $\mathrm{Circ}(X)$ and the sets W and X are disjoint. It is easy to see that the hanging vertices of reverse deduction trees are associated with primitive formulas.

LEMMA 1.4. *A primitive formula is true if and only if it is an axiom in the $F - L - MVD$ calculus.*

Indeed, if the primitive formula $\Sigma \Rightarrow (W \stackrel{*}{\to} Z)$ is not an axiom, then in the interpretation

$$\frac{A_1 \quad A_2 \quad \ldots \quad A_i \quad \ldots \quad A_k}{\begin{array}{ccccc} a_1 & a_2 & \ldots & a_i & \ldots & a_k \\ b_1 & b_2 & \ldots & b_i & \ldots & b_k \end{array}},$$

where $a_i = b_i$ if $A_i \in W$ and $a_i \neq b_i$ if $A_i \notin W$, and all the dependencies from Σ are satisfied but $W \stackrel{*}{\to} Z$ is not satisfied.

COROLLARY 1.5. *A formula $\Sigma \Rightarrow (X \xrightarrow{*} Z)$ is deducible if and only if all the formulas in the list* (3) *are axioms.*

COROLLARY 1.6 (Completeness theorem). *A formula $\Sigma \Rightarrow (X \xrightarrow{*} Z)$ is true if and only if it is deducible in the $F - L - MVD$ calculus.*

Denote by Σ_H the system of dependencies obtained from Σ by deleting all occurrences of attributes from H^c.

THEOREM 1.7. *The rule*
$$\frac{\Sigma_H \Rightarrow (W \xrightarrow{*} Z)}{\Sigma \Rightarrow (W \xrightarrow{*} Z)} \quad (W^c \subseteq H)$$
is semantically correct and reversible.

A *complete axiomatics* for functional, implicit, and multivalued dependencies is also given by the "big step" calculus $F - L - MVD'$ with the same axioms and the following inference rules

Introduction of FD:
$$\frac{\Sigma \Rightarrow (WRY \xrightarrow{*} Z)}{\Sigma, (R \to Y) \Rightarrow (WR \xrightarrow{*} Z)}.$$

Introduction of LD:
$$\frac{\Sigma \Rightarrow (WRA \xrightarrow{*} Z)}{\Sigma, \mathrm{Circ}(RA) \Rightarrow (WR \xrightarrow{*} Z)}.$$

Introduction of MVD:
$$\frac{\Sigma \Rightarrow (WRY \xrightarrow{*} Z) \; \Sigma \Rightarrow (WRY^c \xrightarrow{*} Z)}{\Sigma, (R \to\to Y) \Rightarrow (WR \xrightarrow{*} V)}.$$

This calculus also turns out to be a *lossless* one: all inference rules are correct and reversible, reverse deduction trees have no dead-ends, and Corollary 1.5 is satisfied.

Consider the following *algorithm ρ recognizing whether a formula $\Sigma \to (X \xrightarrow{*} Z)$ is true*:
1. Construct a reverse deduction tree for $\Sigma \Rightarrow (X \xrightarrow{*} Z)$.
2. If all the formulas in the hanging vertices of this tree are axioms, then output the affirmative answer. Otherwise the answer is negative.

COROLLARY 1.8. *The algorithm ρ is correct.*

Let us estimate the run time of ρ. Clearly, the effectiveness of ρ is proportional to the size of the reverse deduction trees being constructed. The only rule that introduces branching is the rule of introducing MVD. However, in this case empty branchings are possible. A "reverse" application of the rule of introducing MVD will be called *trivial* if $WY = U$ or $WY^c = U$. In this case one of the two formulas $\Sigma \Rightarrow (WY \xrightarrow{*} Z)$ or $\Sigma \Rightarrow (WY^c \xrightarrow{*} Z)$ is an axiom and, therefore, the corresponding branch is no longer of interest. We can assume that all such trivial branchings in reverse deduction trees are eliminated.

THEOREM 1.9. *The algorithm ρ is a polynomial one. Furthermore, the run time of ρ does not exceed $O(|\Sigma| \cdot k)$ (here $|\Sigma|$ denotes the record length, i.e., the number of occurrences of attributes in the system of dependencies Σ).*

PROOF. In an arbitrary reverse deduction tree, any branch is passed in no more than $|\Sigma|$ steps, because the reverse application of nonbranching rules always reduces the length of the formula. An evident upper bound for the run time is $\rho - |\Sigma| \cdot p$, where p is the number of hanging vertices in the reverse deduction tree, or the number of nontrivial reverse applications of the rule of introducing MVD. It only remains to show that $p \leq k$.

A deduction in the $F - L - MVD$ calculus provides *additional interesting information* about the basis of dependencies. Following [1], we shall now formulate the definition of the basis of dependencies (a stronger one than in [1], to include both the functional and multivalued cases).

A system of pairwise disjoint nonempty sets

$$W_0; W_1, W_2, \ldots, W_p$$

will be called a *basis of dependencies for X with respect to Σ* if
1. a functional dependency $X \to Z$ logically follows from Σ if and only if $Z \subseteq W_0$,
2. a multivalued dependency $X \twoheadrightarrow Z$ logically follows from Σ if and only if $Z \setminus W_0$ is the union of some set from the list $W_1, W_2, \ldots, W_p$.

Denote the integer p in this definition by $p(X, \Sigma)$, and the cardinal number of the set W_0 by $q(X, \Sigma)$. If $Z \setminus W_0$ is the union of m sets from the list $W_1, W_2, \ldots, W_p$, then m will be denoted by $p(X, Z, \Sigma)$.

LEMMA 1.10. *The following bounds hold:*

$$p(X, Z, \Sigma) \leq p(X, \Sigma), \qquad p(X, \Sigma) + q(X, \Sigma) \leq k.$$

THEOREM 1.11. *For any reverse deduction tree for $\Sigma \Rightarrow (X \stackrel{*}{\to} u)$, if*

(3) $$\Sigma_1 \Rightarrow (X_1 \stackrel{*}{\to} u), \Sigma_2 \Rightarrow (X_2 \stackrel{*}{\to} u), \ldots, \Sigma_p \Rightarrow (X_p \stackrel{*}{\to} u)$$

is the list of formulas located at its hanging vertices, then the system of nonempty sets from the list

$$W_0 = \bigcap_i X_i; \quad W_1 = X_2^c, W_2 = X_2^c, \ldots, W_p = {}_p^c,$$

forms a basis of dependencies for X with respect to Σ.

PROOF. (1) The formula $\Sigma \Rightarrow (X \to Z)$ is true if and only if all $\Sigma_i \Rightarrow (X_i \to Z)$ are axioms, or if $\forall i(Z \subseteq X_i)$, or if $Z \subseteq W_p$.

(2) The formula $\Sigma \Rightarrow (X \twoheadrightarrow Z)$ is true if and only if all $\Sigma_i \Rightarrow (X_i \twoheadrightarrow Z)$ are axioms, or if $Z \subseteq X_i \vee Z^c \subseteq X_i$), which is equivalent to $(X_i^c \cap Z = \emptyset \vee X_i^c \subseteq Z)$.

Note that the proof makes use of the fact that inference rules are paired: the reverse deduction tree is invariant with respect to the choice the metasymbol $\stackrel{*}{\to}$.

Let us estimate the size of the reverse deduction tree.

THEOREM 1.12. *The number of nontrivial applications of the rule of introducing MVD in an arbitrary reverse deduction tree for $\Sigma \Rightarrow (X \stackrel{*}{\to} Z)$ does not exceed $p(X, Z, \Sigma) + q(X, \Sigma) - |X|$.*

PROOF. A reverse application of nonbranching rules does not reduce the main premise WA, while for branching rules we have $WY \cup WY^c = U$. Thus, by virtue of Theorem 1.11 the number of hanging vertices does not exceed $p(X, Z, \Sigma) + q(X, \Sigma) - |X|$.

Theorem 1.12 and Lemma 1.10 imply the relation $p \leq k$ required in the proof of Theorem 1.9.

Theorem 1.11 suggests the following *algorithm α constructing the basis of dependencies for X with respect to Σ.*

(1) Construct any reverse deduction tree for $\Sigma \Rightarrow (X \to U)$.

(2) Using the list of formulas in hanging vertices (3), output as the basis of dependencies for X with respect to Σ the list of all nonempty sets from the list

$$W_0 = \bigcap X_i; \quad W_1 = X_1^c, W_2 = X_2^c, \ldots, W_p = X_p^c.$$

Theorem 1.11 immediately implies the correctness of the algorithm α.

COROLLARY 1.13. *The algorithm α is correct.*

Theorem 1.12 allows us to obtain a quadratic bound for the run time of α.

COROLLARY 1.14. *The algorithm α is polynomial. Furthermore, the run time of α does not exceed $O(|\Sigma| \cdot k)$.*

In fact the algorithms ρ and α work even faster. In order to obtain better bounds, consider their performance in more detail.

The algorithm α (and ρ) can be interpreted as an algorithm constructing a tree and traversing it in depth. Each step along the tree is connected with a reverse application of inference rules and requires *local* modifications with respect to a single letter (the letter A). Under an *appropriate organization of the structure* of the formula $\Sigma \Rightarrow D$ (for example, in the form of 2-connected lists, etc.) such a step can be realized in *constant time*.

In order to reduce the size of the reverse deduction tree and eliminate redundancies we shall observe the following *deduction search strategy*:

(1) Before making a nontrivial application the rule of introducing MVD, one should, as long as possible, apply "in reverse" all other rules.

(2) Instead of the rule of introducing MVD, use the introduction rule

$$\frac{\Sigma_{H'} \Rightarrow (WY \stackrel{*}{\to} Z) \quad \Sigma_{H''} \Rightarrow (WY^c \stackrel{*}{\to} Z)}{\Sigma, (\to \to Y) \Rightarrow (W \stackrel{*}{\to} Z)},$$

where H' and H'' are arbitrary lists of attributes such that

$$(WY)^c \subseteq H', \quad (WY)^c \subseteq H'',$$

and if $WY = U$, then the left premise in the rule is dropped, and if $WY^c = U$, then the right premise is dropped; the fact that this rule is correct and reversible follows from Theorem 1.7.

(3) If the branching is nontrivial, the first (from the left) subtree being traversed is the one whose main premise is "shorter".

(4) When going from the left subtree to the right subtree, the results obtained in traversing the left subtree are used without recalculations.

We shall describe this strategy in precise terms in the form of the recursive *procedure CIRCUIT(v, H)*.

Let v be a nontrivial branching vertex with respect to the rule of introducing MVD at which the formula

$$\Sigma, (\rightarrow\rightarrow Y) \Rightarrow (W \stackrel{*}{\rightarrow} Z)$$

is located, and let $H = W^c$. Let $H_1 = Y \setminus W$ and $H_2 = Y^c - W$.

Note that $WY = WH_1$, $WY^c = WH_2$, $H_1 \cup H_2 = H$, $H_1 \cap H_2 = \emptyset$.

Suppose that there is a preference condition $Q(H_1, H_2)$.

The *process CIRCUIT(v, H)* goes on as follows:

ascend to vertex v;

if $Q(H_1, H_2)$, then

(1) apply "in reverse" nonbranching rules to the formula $\Sigma \Rightarrow (WH_1 \stackrel{*}{\rightarrow} Z)$ as long as it is possible,

(2) let v_1 be the last vertex in the linear portion of the resulting left subtree with the formula $\Sigma' \Rightarrow (WH_1 H_2' \stackrel{*}{\rightarrow} Z)$, where $H_2' \subseteq H_2$, then execute the process $CIRCUIT(v_1, H_2 \setminus H_2')$ in order to obtain the reverse deduction tree for this formula,

(3) return from v_1 to v_2, restore the system Σ_{H_1}, and start moving from v to the right applying nonbranching rules to the formula $\Sigma_{H_1} \Rightarrow (WH_2 \stackrel{*}{\rightarrow} Z)$ as long as possible,

(4) let v_2 be the last vertex in the linear portion of the resulting right subtree with the formula $\Sigma'' \Rightarrow (WH_2 H_1' \stackrel{*}{\rightarrow} Z)$, where $H_1' \subseteq H_1$, then execute the process $CIRCUIT(v_2, H_1 \setminus H_1')$ in order to obtain the right subtree;

if $Q(H_1, H_2)$ is not satisfied, then swap H_1 with H_2 in the rules (1)–(4);

descend from vertex v.

The *reduction in run time* of the algorithm α is due to the *asymmetry* of the $CIRCUIT$ procedure: in the direct and reverse passage through the left segment $v - v_1$ occurrences of attributes from H_1 are affected at most twice, occurrences of attributes from H_2' are affected once and then "leave the game" completely; the direct and reverse passages through the right section $v - v_2$ affect only attributes from H_1' which completely "leave the game", but do not affect attributes from H_2.

A *left shoot* of the branch ξ is the longest linear segment whose first arc is of the form (v, v'), where v lies on the branch ξ, v' is the left son of v, but v' does not lie on the branch ξ. The number of all different left shoots of the branch ξ will be called the *left height* $h\xi$ of this branch. The *left tree height* is the maximal left height of its branches.

An analysis of the $CIRCUIT$ procedure yields the following statements.

THEOREM 1.15. *A given occurrence of the attribute A is affected (except, maybe, one other time) only on the left shoots of the same branch ξ. Therefore, the number of times the occurrence A is affected does not exceed $2h\xi + 1$.*

COROLLARY 1.16. *The run time of algorithm α does not exceed $O(|\Sigma| \cdot h)$, where h is the left height of the reverse deduction tree.*

Since the left height for an arbitrary reverse deduction tree is bounded by the number of multivalued dependencies in Σ, we immediately obtain the following bound.

COROLLARY 1.17. *The run time of the algorithm α does not exceed $O(|\Sigma| \cdot (\mu + 1))$, where μ is the number of multivalued dependencies in Σ.*

If in the $CIRCUIT$ procedure we use a preference condition $Q(H_1, H_2)$ of the form

$$|H_1| \leq c|H_2|, \tag{4}$$

where c is a positive constant, then the left height of the resulting reverse deduction tree for $\Sigma \Rightarrow (X \xrightarrow{*} Z)$ is bounded by $O(\log_2 k)$. This observation substantially lowers the bound from Theorem 1.9 and Corollary 1.14.

COROLLARY 1.18. *The run time of the algorithm α that utilizes in its CIRCUIT procedure a preference condition of the form (4) does not exceed $O(|\Sigma| \cdot \log_2 k)$.*

Another way to obtain a good bound is based on the direct computation of the number of times an attribute occurrence is affected when passing through the reverse deduction tree in the $CIRCUIT$ procedure.

The main part of the run time taken to pass v-v_1-v-v_2-v is proportional to $|H_1|$, since the sum of all $|H_1'|$ over right shoots does not exceed $|\Sigma|$. Let us contract all linear segments of the reverse deduction tree and label the amalgamated vertices v, v_1, and v_2 by the numbers $|\Sigma_U|$, $|\Sigma_{H' \setminus H_2'}|$, and $\Sigma_{H_1 \setminus H_1'}$ respectively. Thus, we arrive at the problem of estimating the sum of all n_v over v that are right sons for a binary tree with labeled vertices $v \to n_v$.

THEOREM 1.19. *Suppose that vertices v in a binary tree are labeled by the positive numbers n_v in such a way that if v' and v'' are the left and the right sons of v, then $n_{v'} + n_{v''} \leq n_v$ and $n_{v''} \leq \gamma n_v$, where γ is the positive root of the equation*

$$\gamma = -\gamma \log_2 \gamma - (1-\gamma) \log_2(1-\gamma).$$

Then the sum of all $n_{v''}$ over v that are right sons does not exceed

$$n_{v_0} \cdot \log_2(p+1),$$

where v_0 is the root of the tree, and p is the number of its vertices.

The proof is achieved by an easy induction. Incidentally, the number $\gamma \approx 0.77$ is found *exactly*. I have a counterexample for a bigger γ.

COROLLARY 1.20. *The run time for the algorithm α whose CIRCUIT procedure uses the preference condition $Q(H_1, H_2)$ of the form*

$$|\Sigma_{H_1}| \leq c \cdot |\Sigma_{H_2}|, \tag{5}$$

where c is a constant such that $\frac{1-\gamma}{\gamma} \leq c \leq \frac{\gamma}{1-\gamma}$, does not exceed

$$O(|\Sigma| \cdot \log_2(p(X, \Sigma) + q(X, \Sigma) - |X|)).$$

The run time bounds for the algorithm α also hold for the algorithm ρ. One can make the algorithm ρ work faster by making use of the fact that inference rules are indifferent to Z and terminating the process of constructing the reverse deduction tree for $\Sigma \Rightarrow (A \stackrel{*}{\to} Z)$ at the vertices where axioms appear. In particular, such algorithms ρ have the following run time bound.

COROLLARY 1.21. *The run time of the algorithm ρ whose CIRCUIT procedure makes use of the preference condition (5) does not exceed*

$$O(|\Sigma| \cdot \log_2(p(X, Z, \Sigma) + q(X, \Sigma) - |X|)).$$

THEOREM 1.22. *The run time on the formula $\Sigma \Rightarrow (X \stackrel{*}{\to} Z)$ of the algorithm ρ whose CIRCUIT procedure makes use of the preference condition $Q(H_1, H_2)$ of the form*

(6) $$|H_1 \cap Z| \leq c \cdot |H_2 \cap Z|,$$

where c is a positive constant, does not exceed

$$O(|\Sigma| \cdot \log_2 |Z - X|).$$

If $|Z| = 1$, then the reverse deduction tree becomes a linear one.

THEOREM 1.23. *The run time of the algorithm ρ on the formula $\Sigma \Rightarrow (X \stackrel{*}{\to} A)$ does not exceed $O(|\Sigma|)$.*

An effective recognition of derived implicit dependencies is done in the *extended calculus* S' obtained from the $F - L - MVD$ calculus by adding another inference rule.

Introduction of LD on the right:

$$\frac{\Sigma \Rightarrow (X_1 \to B_1) \; \Sigma \Rightarrow (X_2 \to B_2) \; \ldots \; \Sigma \Rightarrow (X_m \to B_m)}{\Sigma \Rightarrow \mathrm{Circ}(X)},$$

where $X = B_1 B_2 \ldots B_m$ is an m-tuple of attributes, $X_i = X \setminus \{B_i\}$.

This inference rule is also semantically *correct and reversible*. It is not hard to see that the reverse deduction trees constructed for the formulas $\Sigma \Rightarrow \mathrm{Circ}$ in the extended calculus S' have a single nontrivial branching point at the root of the tree and contain $|X|$ branches.

THEOREM 1.24. *The run time of the algorithm ρ recognizing whether a formula $\Sigma \Rightarrow (X)$ is true by constructing reverse deduction trees in the calculus S' does not exceed $O(|\Sigma| \cdot |X|)$.*

The effectiveness of the algorithms α and ρ does not decrease if we consider the *complete hierarchical dependencies* introduced in [3].

A complete hierarchical dependency

$$X \to Z_1 | Z_2 | \ldots | Z_n |$$

is *satisfied* in the interpretation $r(U)$ if the implication

$$(\exists Y_1 r(U) \; \& \; \exists Y_2 r(U) \; \& \; \ldots \; \& \; \exists Y_{m+1} r(U) \supset r(U)$$

is an identity, where $Y_i = Z_i^c \setminus X \; (i = 1, 2, \ldots, m), Y_{m+1} = (\bigcap Z_i) \setminus X$.

Using an appropriate natural extension of the calculus S', we obtain the algorithms α and ρ with the bounds from the preceding theorems. In particular, let us give bounds for the analysis algorithm ρ.

THEOREM 1.25. *Let Σ be a system of functional, implicit, multivalued, and complete hierarchical dependencies, μ the number of multivalued and complete hierarchical dependencies in Σ. Then:*
1. *The run time of the algorithm ρ on the formula $\Sigma \Rightarrow (X \to Z)$ does not exceed*
$$(O(|\Sigma| \cdot \min(\mu + 1, \log_2 |\Sigma \setminus X|)).$$
2. *The run time of ρ on $\Sigma \Rightarrow (X \stackrel{*}{\to} B)$ does not exceed $O(\Sigma)$.*
3. *The run time of ρ on the formula $\Sigma \Rightarrow \mathrm{Circ}(X)$ does not exceed $O(|\Sigma| \cdot |X|)$.*
4. *The run time of ρ on $\Sigma \Rightarrow (X \to |Z_1| \ldots |Z_m|)$ does not exceed*
$$O(|\Sigma| \cdot \min(\mu + 1), \min_j \log_2 |Z_j \setminus X|)).$$

2. Schematic synthesis of minimal programs

In this section we use lossless calculi to obtain effective algorithms for the synthesis of certain types of programs. For simplicity we shall restrict our attention to the following problem.

Let Σ be a system of functional and implicit dependencies. Each dependency is labeled by the name of a function, procedure, program module, etc., realizing it.

By a *program* we mean a linearly ordered list of dependency names $\pi; f_1; \ldots; f_l$. The corresponding translation to a desired programming language is evident.

A *computation of the program π at the output X* is defined by induction:
Step 0. $X^0 = X$.
Step t. If f_t is the name of the functional dependency $X_t \to Y_t$ and $X_t \subseteq X^{t-1}$, i.e., "all variables X_i" are computed, then $X^t = X^{t-1} \cup X_i$; in other cases X^t is not defined.

A *program π solves the problem $X \to Z$* if $Z \subset X^l$ "for given X computes all variables Z". For details see, e.g., [2].

The *synthesis problem* is stated as follows: given a system Σ and a "problem" $X \to Z$, find a program π solving the problem $X \to Z$.

In order to solve the synthesis problem, we shall use the $F - LD$ calculus obtained from the $F - L - MVD$ calculus by deleting the axioms and inference rules involving multivalued dependencies.

THEOREM 2.1. *There is a correspondence, computable in real time, between programs solving the problem $X \to Z$ and deductions of the formula $\Sigma \Rightarrow (X \to Z)$ in the $F - LD$ calculus.*

PROOF. From the program solving the problem $X \to Z$ one can uniquely reconstruct a deduction of the formula $\Sigma \Rightarrow (X \to Z)$, because step t of the computation of π on X can be represented as the transition from $\Sigma_{t-1} \Rightarrow (X^{t-1} \to Z)$ to $\Sigma_t \Rightarrow (X^t \to Z)$, which is evidently modeled in the $F - LD$ calculus by a reverse application of several weakened rules and one introduction rule (here Σ_t is the system Σ from which the dependencies $f_1, \ldots, f_t$ are deleted).

On the other hand, by writing in reverse order the names of dependencies appearing in the application of the rules of introducing FD and LD in an arbitrary deduction of the formula $\Sigma \Rightarrow (X \to Z)$, we obtain the program solving the problem $X \to Z$.

Note that if in the course of constructing the reverse deduction tree we write down the names of "disappearing" dependencies, we obtain the desired "direct" program.

In the absence of multivalued dependencies the reverse deduction tree does not branch at all (*there is no search*). The following corollary is evident.

COROLLARY 2.2. *The synthesis problem based on the $F-LD$ calculus is solved in time $O(|\Sigma|)$.*

Unfortunately, programs that can be synthesized in linear time may turn out to be too cumbersome.

Consider two *criteria for evaluating the program quality*.

The *run time* of the program π is the sum of run time bounds for each module f_t. In the simplest case the run time of the program π can be understood as the number of executable modules l.

A program with minimal run time corresponds to a minimal deduction.

THEOREM 2.3. *For any propositional formula F in n variables one can construct, in polynomial time, a system of functional dependencies Σ such that*
1. *$|\Sigma| = O(n \cdot |F|)$, where $|F|$ is the number of logical connectives in F,*
2. *all dependencies in Σ are of the form $X \to Y$ satisfying the conditions: $|X| \leq 2$, $|Y| = 1$ (such dependencies can be called* binary),
3. *the dependency $A_1 \to A_2$ logically follows from the system Σ,*
4. *if formula F is satisfiable, then the minimal length of the deduction $\Sigma \Rightarrow (A_1 \to A_1)$ does not exceed $(2n+3)|F|$, and the minimal length (the number of executable modules) of the program solving the problem $A_1 \to A_2$ does not exceed $(n+1)|F|$,*
5. *if F is unsatisfiable, then the minimal length of the deduction for $\Sigma \Rightarrow (A_1 \to A_2)$ is no smaller than $(2n+4)|F|$, and the minimal length of the program solving the problem $A_1 \to A_2$ is no smaller than $(n+2)|F|$.*

PROOF. Associate with each subformula G_0 of the formula F the pair of attributes G_0 ("G is false") and G_{s_1} ("G is true").

Write out the following functional dependencies. If $G_0 = G_0' \mathbin{\&} G_0''$, then write

$$G_0' \to G_0, \ G_0'' \to G_0, \ G_1'G_1'' \to G_1.$$

If $G_0 = G_0' \vee G_0''$, then write

$$G_0'G_0'' \to G_0, \ G_1' \to G_1, \ G_1'' \to G_1.$$

For $G_0 = \neg G'$ write $G_1' \to G_1, \ G_1' \to G_0$.

With each variable x_i we have already associated the attributes X_{i0} (x_i is *false*) and X_{i1} (x_i is *true*). Construct two chains of dependencies from $|F|$ which imply $A_1 \to X_{i0}$ and $A_1 \to X_{i1}$. Finally, add the dependency $F_1 \to A_2$. Also, construct a *long* chain of dependencies which implies $A_1 \to A_2$.

It is not hard to see that if F is satisfied by the collection $(\gamma_1, \gamma_2, \ldots, \gamma_n)$, then $A_1 \to A_2$ can be realized as follows: first, move from A_1 to $X_{1\gamma_1}, X_{2\gamma_2}, \ldots, X_{n\gamma_n}$, and then apply dependencies corresponding to the consecutive steps of computing the value of formula F on this collection. The length of such a program does not exceed $(n+1)|F|$.

If F is unsatisfiable, then in any program solving the problem $A_1 \to F_1$ for at least one i both X_{i0} and X_{i1} must be computed. Therefore the length of such a program is no less than $(n+1)|F| + |F|$. The remaining details are evident.

COROLLARY 2.4. *The synthesis problem for programs with minimal run time is NP-complete (even in the class of systems of binary dependencies).*

The difficulty of the problem is not reduced if we search for not necessarily minimal programs but programs whose run time is *close to minimal*. Generalizing Theorem 2.3 we obtain the following bound for the complexity of almost minimal programs.

THEOREM 2.5. *The synthesis problem for programs whose run time differs from minimal by at most $|\Sigma|^{1-\varepsilon}$ is NP-complete (in the class of systems of binary dependencies).*

The program complexity can be estimated by the size of its expansion.

The *expansion of the solution of the problem $X \to B$ with respect to Σ* is a tree whose vertices are labeled by variables (attributes) such that
1. all hanging vertices are labeled by the input variables (variables from X),
2. the root is labeled by the output variable B,
3. every nonhanging vertex v with label A is marked by the name of the dependency f from Σ in such a way that if f is a functional dependency $W \to Y$, then $A \in Y$, $W \subseteq W_v$, where W_v is the set of all variables labelling the sons of v; if f is an implicit dependency $\mathrm{Circ}(W)$, then $W \subseteq W_v A$.

It is clear that any program solving the problem $X \to Z$ can be expanded. The expansion corresponds to the representation of the program in the form of a *functional term*.

With the help of an appropriate strategy for deduction search in the $F - LD$ calculus it is possible to obtain a polynomial synthesis algorithm for programs with expansions of minimal size.

THEOREM 2.6. *The synthesis program for programs with expansions of minimal size can be solved in time $O(|\Sigma| \cdot \log_2 k)$, where k is the number of vertices.*

PROOF. In the process of reverse deduction each variable A and each dependency f is associated with the current complexity of its expansion $\nu(A)$ and $\nu(f)$ respectively. All the variables A are linearly ordered by $\nu(A)$.

First, at step 0, we set $\nu(X) = 0$ for input variables X, $\nu(A) = w$ for the remaining A, and $\nu(f) = 0$ for all f.

The ordinary step of the reverse deduction is as follows. Let C be the first variable in the linearly ordered list. Perform all possible reverse applications of the weakened rule with respect to C. For each dependency f affected by the process, increase $\nu(f)$ by $\nu(C)$. Next, if the dependency $g: (\to Y)$ appears in Σ, apply the reverse rule of introducing FD, and for any A in Y set $\nu(A)$ equal to the minimum of the old value of $\nu(A)$ and the number $\nu(g)+1$. If the new dependency is $g\colon\mathrm{Circ}(A)$, apply the reverse rule of introducing LD, and set $\nu(A)$ equal to the minimum of the old value of $\nu(A)$ and the number $\nu(g)+1$. The list of variables is then reordered by the new values of ν. "Moving" every variable into its right place requires *at most* $\log_2 k$ steps. Finally, delete the variable C from this list and from further consideration.

It can be shown that the stabilizing value of $\nu(B)$ coincides with the size of the minimal expansion for the solution of the problem $X \to B$, while the minimal expansion program can be constructed from the deduction obtained by the above procedure until the moment when the variable B was deleted.

Since processing the succeeding occurrence of the variable C requires at most $\log_2 k$ steps, we obtain the desirable run time bound for the synthesis algorithm for minimal expansion programs.

Note that similar results are obtained in determining program and expansion complexity taking into account the *system of positive weights for initial modules*:
1. the synthesis problem for programs with minimal run time is NP-complete (even in the class of binary dependencies);
2. the synthesis problem for programs with minimal expansions is solved in time $O(|\Sigma| \cdot \log_2 k)$.

For example, let us synthesize a program computing $x \to x^7$ using one-time multiplication modules, i.e., dependencies of the form

$$x \to x^2, xx^2 \to x^3, \ldots, xx^6 \to x^7, \ldots$$
$$x^2 \to x^4, x^2x^3 \to x^5, \ldots, x^3 \to x^6, x^3x^4 \to x^7, \ldots .$$

The algorithm of Theorem 2.6 yields the following *minimal expansion program*:

$$f_1\colon x \to x^2;$$
$$f_2\colon xx^2 \to x^3;$$
$$f_3\colon x^3 \to x^6;$$
$$f_4\colon xx^6 \to x^7,$$

or, in the form of the functional term,

$$f_4(x, f_3(f_2(x(f_1(x))))).$$

The idea of the algorithm from Theorem 2.6 makes it possible to construct the *synthesis algorithm for programs with minimal run time*, which works fast in "practical" and uncomplicated situations: the roles of $\nu(A)$ and $\nu(f)$ are played by collections of programs solving the problems of the type $X \to A$ and $X \to W$ respectively; at the ordinary step of the reverse deduction in the new collection $\nu(f)$ for the affected dependency F we take into account all versions of computing C from X. In this case it is helpful to use the a priori bound on the size of the program from Corollary 2.2.

In our example this algorithm yields two *minimal run time* programs for the problem $x \to x^{15}$:

$$\begin{array}{ll} x \to x^2; & x \to x^2; \\ xx^2 \to x^3; & xx^2 \to x^3; \\ x^2x^3 \to x^5; & x^3 \to x^6; \\ x^5 \to x^{10}; & x^3x^6 \to x^9; \\ x^5x^{10} \to x^{15}; & x^6x^9 \to x^{15}. \end{array}$$

We conclude by noting that in the case of *unary* dependencies $X \to Y$ for which $|X| = |Y| = 1$, both criteria for program complexity bounds coincide. As a result, we obtain, in particular, *an algorithm with run time $O(|E| \cdot \log_2 |V|)$ searching for a shortest path between given vertices* in an oriented graph ($|V|$ is the number of vertices, $|E|$ is the number of arcs).

References

1. J. Ullman, *Principles of database systems*, Computer Science Press, Potomac, MD, 1980.
2. G. E. Mints and E. H. Tyugu, *Structural synthesis and nonclassic logics*, Abstracts of the 3rd Conference "Application of Methods of Mathematical Logic", Tallinn, 1983, pp. 52–60. (Russian)
3. C. Delobel, *Normalization and hierarchical dependencies in the relational data model*, ACM Trans. Database Systems **3** (1978), 201–222.
4. Zvi Galil, *An almost linear algorithm for computing a dependency basis in a relational database*, Lecture Notes in Computer Sci., vol. 85, Springer-Verlag, Berlin, 1980, pp. 246–256.
5. E. Ya. Dantsin, in this volume.
6. A. Ya. Dikovsky, *Effectiveness bounds for algorithms related to computational models*, Abstracts of the 3rd Conference "Application of Methods of Mathematical Logic", Tallinn, 1983, pp. 42–51. (Russian)
7. I. O. Babaev, S. S. Lavrov, G. A. Netsvetaeva, F. A. Novikova, and G. M. Shuvalov, *SPORA, a programming system with automatic program synthesis*, Abstracts of the 3rd Conference "Application of Methods of Mathematical Logic", Tallinn, 1983, pp. 29-41. (Russian)
8. O. Yu. Gorchinskaya, *Theoretical aspects of constructing relational databases*, Avtomat. i Telemekh. **1983**, no. 1, 5–25; English transl., Automat. Remote Control **44** (1983), no. 1, 1–17.
9. M. Garey and D. Johnson, *Computers and intractability: a guide to the theory of NP-completeness*, Freeman, San Francisco, CA, 1979.
10. A. Aho, J. Hopcroft, and J. Ullman, *The design and analysis of computer algorithms*, Addison-Wesley, Reading, MA, 1976.

Lower Bounds of Combinatorial Complexity for Exponential Search Reduction

N. K. Kossovskiĭ

ABSTRACT. The paper presents a simple example of a sequence of Boolean functions that does not allow a schematic realization with the number of elements expressed by an exponential function (with some base) of the arity of the Boolean function. The example is based on a problem which can be solved on a nondeterministic Turing machine in a number of steps expressed by a double exponent. This example has the maximal order of combinatorial complexity. Any two Boolean functions of the proposed sequence coincide if the extra arguments of the function which has more variables are set to zero.

In the definition of a sequence of Boolean functions no use is made of the notion of complexity or universal function in any form. The first example admitting a schematic realization with only an exponential number of elements, is due to L. J. Stockmeyer, 1974. This example is described in detail in Chapter 7, Section 2 of the book by R. G. Nigmatullin [1], which also contains a sufficiently detailed history of the question and includes a bibliography. Unfortunately, the formal theory that is close to arithmetic and usable in the example of the given type is algorithmically unsolvable. The example suggested in this article is based on establishing the solvability of logical-arithmetic equations with exponential bounds on the length of solutions. In fact, the result earlier obtained by the author about lower bounds for establishing solvability of such equations by means of Turing machines is transferred to schemes of functional elements.

Computations by schemes of functional elements are modeled by Turing machines computations in a bounded number of steps. As was proved in [2], these machines can be represented by logical-arithmetic equations with respect to short solutions. By a logical-arithmetic equation we mean an equality of polynomials with positive integer coefficients with extra additive terms that are the results of Boolean multiplication of binary numbers (logical multiplication used in computers). The solvability of equations is considered with respect to short solutions.

THEOREM. *There exists a rational number C greater than 1 such that for any positive integer M the establishment of solvability for logical-arithmetic equations with a bound on solution length of the form $2^{2^{ME}}$, where E is a positive integer,*

©1996 American Mathematical Society

cannot be realized by a scheme of C^M functional elements—Sheffer's symbols. Here the lengths of the equations subject to realization are bounded from above and below by linear functions of M.

In a schematic realization of logical-arithmetic equations of bounded length all the letters of the alphabet in which the equations are written are coded by binary words of fixed length. Each bit is then transmitted to the input of the functional scheme. In order that all the equations have the same length an extra letter is used, coded by the word consisting only of zeros.

The importance of the problem of constructing individual sequences of Boolean functions having high complexity of realization in the class of schemes of functional elements was noted by O. B. Lupanov [3] and S. V. Yablonskiĭ [4]. The proposed example of the sequence of Boolean functions can be compared with the result of L. A. Sholomov, 1975 [5].

It is important that the proof of the above theorem makes triple use of reflexive (diagonal-cardinal) constructions. For example, the proof of the theorem from [2] uses such constructions twice.

Essentially, the proposed sequence of Boolean functions approximates some decidable predicate. R. I. Freidzon [6] proved that, except for a set of measure zero, all recursive predicates have complex approximations. However, up to now there are no examples of simple nonartificial decidable predicates of low complexity (for example, decidable in an exponential number of steps on nondeterministic Turing machines).

Note that in logical-arithmetic equations the operation of bitwise multiplication of binary representations of operands is used. It is an important operation performed in computers and can be used effectively for rather long operations. For example, this and other bitwise operations were used by V. L. Temov and M. B. Matskin [7] for a sufficiently fast solution of a search problem of graph coloring on the ES-1012 computer.

The last part of the article is devoted to the expressibility of nondeterministic calculations by equations over bit strings with solutions of bounded length.

The author is grateful to R. N. Nigmatullin, whose interesting question provoked an answer contained in the first part of the article, as well as to the participants of the seminar conducted by R. I. Freidzon, which made the appearance of the second part of the article possible.

LEMMA 1. *There exists a sequence of Boolean functions such that* (1) *a Boolean function of a smaller number coincides with the Boolean function of the greater number whose additional arguments are equal to zero, and* (2) *starting with some M these functions are nonrealizable by schemes of functional elements whose number does not exceed $2^{M/2}$, where M is the arity of the Boolean function.*

For the sake of simplicity, we assume that Sheffer's function is used for functional elements, and that there are two constant inputs, 0 and 1. The lemma follows easily from a widely known result for schemes of functional elements realizing conjunction, disjunction, and negation. In this case the lower bound obtained by Lupanov [8] is, starting from some M, of the form $2^{M/2}$. It remains to note that for any $C > 1$ starting with some M the expression $2^m/M$ is greater than $2^{M/2}/C$, and that in order to realize Sheffer's symbol in this basis two elements are sufficient. A detailed proof of this result of Lupanov is given by Nigmatullin in [**1**], Chapter 5, Lemmas 8 and 9.

LEMMA 2. *One can construct a one-sided one-tape nondeterministic Turing machine which, given a sequence M of logical constants, computes the value of the scheme of the sequence in Lemma 1 on this collection of logical constants. The number of steps does not exceed* $2^{2^{ME}}$ *for some positive E.*

The desired Turing machine runs as follows. Write the first "complex" Boolean function nondeterministically. Then consecutively scan exponentially many "short" schemes. In the course of this, write the short scheme, and write nondeterministically the vector of constants on which the value of the Boolean function realizable by the "short" scheme and the value of the "complex" Boolean function differ, and then check this difference deterministically (in doing so it is required to compute the value of the "short" scheme on this vector of constants). For each of these stages the number of steps required is expressed by a double exponential function of the number of logical constants in the input.

What follows is based on the following lemma, which is a consequence of the main result obtained by the author in [**2**].

LEMMA 3. *The predicate of computability on the Turing machine from Lemma 2 coincides with the predicate of solvability for the logical-arithmetic equation in positive integers whose lengths are bounded by* $2^{2^{ME}}$ *for some E.*

It is sufficient to prove the theorem for sufficiently large M. The proof will be achieved by *reductio ad absurdum*. Apply Lemma 3. The lengths of the resulting logical-arithmetic equations are bounded from above by a linear function in M. By the assumption, for any rational number $C > 1$ there is a required scheme of functional elements whose number is less than C^M, for infinitely many M. Therefore, for all such M, Lemmas 1, 2, and 3 imply that

$$2^{M/2} < C^{HM+H''}$$

for some H, H'', which is impossible for sufficiently large M and sufficiently small C. Indeed, taking into account that the length of the logical-arithmetic equation is bounded from above by the cube of a linear function in M, the variables $X_1, \ldots, X_M$ can be adjoined to the inputs used to write the logical-arithmetic equation, and then one can use the scheme whose existence was assumed. To all the other inputs, reserved for entering the logical-arithmetic equation, we adjoin the necessary constant inputs. It is convenient to assume that the logical constants in logical-arithmetic equations are encoded by sequences consisting of either ones only, or zeros only.

Modern computers usually use bit strings, i.e., nonempty sequences of zeros and ones. Such strings are often processed by bitwise logical operations considering that 0 and 1 encode the logical constants "false" and "true" respectively. Here it is efficient to use sufficiently long strings (for example, in the paper by Temov and Matskin [**7**]).

In what follows we study the complexity of establishing solvability of equations over bit strings containing the following operations over bit strings: bitwise maximum, bitwise minimum (corresponding to bitwise disjunction and bitwise conjunction preceded by equalizing the lengths of strings by adding zeros to the left of the shorter one), concatenation of strings. The results of applying these operations to the strings X, Y will be denoted, respectively, by $X \vee Y$, $X \wedge Y$, XY. In practical application to solving equations it makes sense to consider as solutions only sufficiently short bit strings. Such equations can be interpreted as requests for query search in a database realized in the form of templates defining elements from some set and specifying which properties from some standard list must be satisfied. (The properties are assumed to be ordered by positive integers.) Similarly to the theorem from [2], an equivalent (up to a quadratic function) description is given for predicates computable on nondeterministic machines with bounded time in terms of solvability of equations over bit strings.

Let us proceed to precise statements. In what follows we consider nondeterministic one-tape Turing machines with potentially infinite (to the left only) tape. A predicate is computable on this machine if for any argument the machine processes it and arrives at a halt state if and only if the predicate on this argument is satisfied. Let $|X|$ denote the length of the bit string X, without the nonsignificant zeros on the left, and let $\Phi(X) \geq X$. We shall assume that the list of variables for bit strings does not exceed some expression if the value of each variable from the list does not exceed this expression in length.

THEOREM. (1) *If for some positive numbers C, C' and any X the predicate Π depending on the argument X is computable by the nondeterministic Turing machine in at most $C' \cdot \Phi(|X|)^C$ steps, then for some C'' this predicate can be represented in the form*

$$(*) \qquad \exists Y (\leq C'' \cdot \Phi(|X|)^{2C})(P(X,Y) = P''(X,Y)),$$

where P, P'' are the results of superposing the bitwise minimum and maximum operations and the operation of concatenation, and Y is the list of variables (an equivalent representation of the predicate Π is obtained if the restriction on the quantifier is removed).

(2) *Each predicate that can be represented in the form $(*)$ is computable from X by a nondeterministic Turing machine in at most $C' \cdot \Phi(|X|)^{4C}$ steps for some C'.*

Note that in what follows we shall omit universal quantifiers with respect to free variables in formulas written in a string.

PROOF OF THE THEOREM. Let us first prove the second part of the theorem. Suppose that a predicate can be represented in the form $(*)$. Its nondeterministic computation can be performed as follows. Nondeterministically write the numbers substituted in place of the list Y. This takes no more than $MC'' \Phi(|X|)^{2C}$ steps, where M is the number of arguments in P, P''. Then check the condition

$$P(X,Y) = P''(X,Y).$$

This checking procedure requires no more than

$$C''(|X| + \Sigma|Y| + |P|)^2$$

steps, where $|P|$ is the length of expression P and $\Sigma|Y|$ is the sum of the lengths of all numbers substituted instead of Y.

Thus, a nondeterministic computation of the predicate turns out to be possible in no more than $C'\Phi(|X|)^{4C}$ steps for some C'.

Now let us prove the first part of the theorem. Suppose that a predicate depending on the argument X is computable by a nondeterministic Turing machine in at most $C'\Phi(|X|)^C$ steps. We obtain a representation for such a predicate in the form $(*)$.

Suppose that the external alphabet of the Turing machine is the alphabet $A = \{b_0, \ldots, b_M\}$; b_0 is considered to be the blank symbol of the Turing machine tape. When new squares are added to the tape, it is supposed that they contain b_0. In the rightmost square of the tape stands the symbol $*$ which can neither be erased, nor written into any other square. The machine can be in one of the states $P_0, \ldots, P_H$, where P_1 is the start-state and P_0 is the halt-state. A command of the Turing machine has one of the following forms:

$$P_K b \Rightarrow P_{K''} b'', \quad P_K \Rightarrow \overrightarrow{P}_{K''}, \quad P_K \Rightarrow \overleftarrow{P}_{K''},$$

where b and b'' are letters in the alphabet $A \cup \{*\}$, $K \geq 1$, $K'' \geq 0$, and an arrow over a state denotes the movement of the head one square in the direction given by the arrow. A configuration (or momentary description) is any word of the form $T\overrightarrow{P}_K T'*$ or $T\overleftarrow{P}_K T'*$, where T and T' are words in the alphabet A. These momentary descriptions correspond to the following sequence on the tape: $TT'*$. The arrow indicates the letter scanned by the Turing machine head.

Since we consider nondeterministic Turing machines, it is convenient to single out that part of the indeterminacy which depends only on the state of the Turing machine and the letter being scanned. This results in nondeterministic subprograms having the left-hand part of the form $P_K b$ (condition for the program to start working) and a number of expressions in the right-hand part (which can replace the left-hand part) having one of the following forms: $P_K b''$; $\overrightarrow{P}_{K''}$; $\overleftarrow{P}_{K''}$ (the number of such expressions can be arbitrarily large).

It will be convenient to replace subprograms of this type with two nondeterministic substitutions in the alphabet $A \cup \{\times\} \cup \{\overrightarrow{P}_0, \ldots, \overrightarrow{P}_0, \overleftarrow{P}_0, \ldots, \overleftarrow{P}_H\}$. The left-hand part of the first substitution is of the form $P_K b$ and the right-hand part contains expressions of the form

$$\overrightarrow{P}_{K''} b''; \quad P_{K''} b; \quad b\overrightarrow{P}_{K''}$$

corresponding to the form of the expressions in the right-hand part of the subprogram. The left-hand part of the second substitution is of the form $\overrightarrow{P}_K^B$ and the right-hand part contains expressions of the form

$$b\overleftarrow{P}_{K''}; \quad \overrightarrow{P}_{K''} b; \quad b P_{K''},$$

also corresponding to the form of the expressions in the right-hand part of the subprogram. Note that these substitutions do not change the length of the word in which the substitution is made.

We assume that the Turing machine starts with the configuration $b_0 \ldots b_0 X$, where X is a bit string filled with the digits b_0, b_1 which denote, respectively, 0 and 1.

Suppose that the machine halts after b steps. This is possible if and only if there exist words Φ, G, K, E, and T such that the length of each of them does not exceed $(|X| + e + 2)(e + 1)$ and the following conditions are satisfied.

(0) The word E is obtained by successively writing several configurations one after another.

(1) $K = \Phi b_0 X \overrightarrow{P}_1$.

(2) Φ is of the form $b_0 \ldots b_0$, which is equivalent to $\Phi b_0 = b_0 \Phi$.

(3) KE is obtained from T as a result of a deterministic sequence of substitutions of two-letter words, consisting each time of different previously unused letters, for the left-hand parts of substitutions which have replaced, as mentioned above, the Turing machine commands.

(4) EG is obtained from T as a result of a nondeterministic sequence of substitutions of the same two-letter words as in (3) for the right-hand parts of substitutions (instead of the same words for which left-hand parts of corresponding substitutions were substituted).

(5) G contains $\overleftarrow{P}_0$ or $\overrightarrow{P}_0$.

(6) BT contains none of the letters $\overrightarrow{P}_0, \ldots, \overrightarrow{P}_H, \overleftarrow{P}_0, \ldots, \overleftarrow{P}_H$.

Indeed, it is sufficient to prove that the existence of T (bounded from above by the corresponding expression) which satisfies conditions (3) and (4) is equivalent to the following condition.

(3″) EG is obtained from KE as a result of simultaneous nondeterministic substitutions of the left-hand parts of substitutions, which, as mentioned above, replaced the Turing machine commands, for one of the right-hand parts of the nondeterministic substitution.

Let us prove this statement. If there exists T for which conditions (3) and (4) are satisfied, then, since all previously unused letters are different, we can replace the successive substitution described in (3) and (4) by an equivalent simultaneous substitution. Then, since T contains no letter defining the state of the Turing machine, the simultaneous substitution from condition (3) can be equivalently replaced by the inverse one. The result of the new substitution and condition (4) imply condition (3″). Now suppose that condition (3″) is satisfied. Then one can take for T the result of the successive substitution in KE replacing the left-hand parts of the substitutions, which, as mentioned above, replaced the commands of the Turing machine with two-letter words consisting of new different letters.

The deterministic substitutions used in (3) can be represented as nondeterministic substitutions of two-letter words consisting of different letters for two-letter words, because substitutions can be done for equal two-letter words. It is important that the substitutions of (4) introducing previously unused letters can be reduced to nondeterministic substitutions of the form

$$X = [Y]^{AB}_{CE \| C''E''}$$

(each occurrence of the word AB in word Y is replaced by one the words CE, $C''E''$). Here the letters A, B are different.

In order to complete the proof it is sufficient to represent the conditions with the bit string operations described in the theorem.

Let $K \geq 2$. By the code of a word P in a K-letter alphabet, whose letters will be referred to by their numbers in the sequence of their occurrences in the list of all letters of the alphabet, we shall mean a system of K bit strings of equal length such that each E bit string for $E \leq K$ is obtained from P by replacing each occurrence of the Eth letter by 1 and the occurrences of the remaining letters by 0. Evidently, the word can be uniquely reconstructed from its code, and the code of the empty word is the system of only empty words.

Let $X_1, \ldots, X_K$ be the code of the word X, and $Y_1, \ldots, Y_K$ the code of the word Y. It easy to see that two words are graphically equal if and only if their codes coincide componentwise, and that the code of the word XY is the system $X_1 Y_1, \ldots, X_K Y_K$. Note that the conjunction and disjunction of bit strings can be reduced to equations over bit strings, as was done in [**9**].

This encoding makes it easy to represent the predicate of substituting one letter in a word for another, the predicate "to be a word consisting of one given letter", the predicate "to be a word containing one of two given letters", the predicate "to be a word that does not contain a given letter", and other similar predicates.

The predicate "the system of bit strings $X_1, \ldots, X_K$ is a code for some word" is expressed as follows. First, for $e \neq e''$ we have $(X_e \wedge X_e)0 = 0(X_e \wedge X_e''), X_e \vee (X_e \wedge X_e'') = X_e$; second, $(X_1 \vee \ldots \vee X_K)1 = 1(X_1 \vee \ldots X_K)$.

Note that instead of the two bitwise logical operations $\wedge, \vee$ one can use $\neg, \vee$ or $\neg, \wedge$, since for any X, Y we have the equalities $X \wedge Y = \neg(\neg X \vee \neg Y)$, $X \vee Y = \neg(\neg X \wedge \neg Y)$, which are bitwise analogs of de Morgan's laws. However, modifying the statement of the theorem in terms of the operations $\neg, \vee$ and the two operations $\neg, \wedge$ will require two extra formulations for the theorem we are now proving.

We now proceed to the essential step in the proof: representation in the desired form of the formula for nondeterministic substitution

$$(**) \qquad X = [Y]^{AB}_{CE \| C'E'} \quad (A \neq B),$$

where X and Y are variables for words, and the remaining letters denote letters in the alphabet under consideration.

This function can be represented with the help of either bitwise conjunction (minimum) or bitwise disjunction (maximum).

For the lower indices y in X and Y we shall sometimes use letters instead of letter numbers. It is convenient to denote by $X(AB), X(CE), X(C'E'), Y(AB), Y(CE)$, and $Y(C'E')$ the expressions $X_A \wedge X_B 0, X_C \wedge X_E 0, X_{C'} \wedge X_{E'} 0, Y_A \wedge Y_b 0, Y_C \wedge Y_E 0$, and $Y_{C'} \wedge Y_{E'} 0$, respectively.

Formula $(**)$ is equivalent to the conjunction of the following statements.

(1) $X_1 \wedge X_2 = Y_1 \wedge Y_2$, i.e., the lengths of the words X and Y coincide.

(2) X does not contain the word AB, i.e., $X(AB)0 = 0X(AB)$.

(3) $Y(AB) \wedge ((X(CE) \wedge X(C'E')) = Y(AB)$, i.e., AB occurs in Y only in those places where either CE or $C'E'$ occurs in X.

(4) For any e not exceeding K we have

$$(Y(AB) \vee Y(AB)0 \vee Y_e 0) = (Y(AB) \vee Y(AB)0 \vee X_e 0),$$

i.e., Y_e coincides with X_e except for those places where AB occurs in Y.

This completes the proof of the theorem

Note that in this theorem it is not possible to exclude bitwise operations, since, as shown in [9], the symmetry predicate cannot then be represented in the form $(*)$.

Equations with bounded solutions over bit strings make it possible to obtain lower bounds of combinatorial complexity in complete analogy with those obtained for logical-arithmetic equations in the first part of the article.

The theorem proved in the second part of this paper can be used to obtain a new proof of the representation of computations by logical-arithmetic equations obtained in [2]. The proof is based on the following relations.

(1) Any X and Y satisfy the equality

$$X + Y = (X \vee Y) + (X \wedge Y).$$

(2) The predicate "X is equal to zero or X can be represented in the form $1\ldots 1$, where the number of ones is greater than zero" can be written in the form

$$(X + 1) \wedge X = 0.$$

(3) The equality $X = (Y \wedge X'')$ holds for any X, Y, and X'' if and only if there exists Y'' such that its binary representation consists only of ones, and its length is greater than 0 and greater than or equal to the length of the binary representation of each of the numbers X, Y, and X'', i.e.,

$$\exists A(Y'' = X + A), \quad \exists A(Y'' = Y + A), \quad \exists A(Y'' = X'' + A);$$

the ones in the binary representation of X cannot be located in the positions where there are no ones in Y (no ones in Y''), i.e.,

$$(Y'' - Y) \wedge X = 0, \quad (Y'' - X'') \wedge X = 0;$$

and, finally, the following conditions hold:

$$Y \geq X, \quad X'' \geq x, \quad (Y - X) \wedge (X'' - X) = 0.$$

(4) For any Y and Y'' the equality

$$Y'' + 1 = 2^{|Y|}$$

holds if and only if the binary representation of Y'' consists only of ones (i.e., $(Y'' + 1) \wedge Y'' = 0$, $Y'' \geq 1$); and the lengths of the binary representations of Y and Y'' are equal, i.e.,

$$\left[\frac{(Y'' - 1)}{2}\right] \vee Y = Y''.$$

In the last equation the bitwise maximum operation can be excluded. It is sufficient to replace this equality by the conjunction of the following two conditions:

$$\exists A(Y'' = Y + A), \quad \exists A(Y = \left[\frac{(Y'' - 1)}{2}\right] + A + 1).$$

(5) For any X, Y, and X'' the concatenation of X and Y is equal to X'' if and only if $X \cdot 2^{|Y|} + Y = X''$.

With the help of these five relations one can obtain the theorem from [2] as a corollary to the preceding theorem. Thus, the present article contains all the basic ideas for the proof of the theorem from [2].

For a detailed proof it is enough to successively eliminate the operation of bitwise maximum in the equation over bit strings, then replace the operation of bitwise minimum by the predicate "the operation of bitwise minimum is equal to zero", and, finally, eliminate the operation of concatenation. The newly added operations are those of addition and multiplication, and each string is defined by a pair of positive integers, of which the first has the string as its binary representation, and the second is the square of the length of the string. Here we assume that both 0 and the empty string are binary representations of the number 0. With the help of the above relations we can express the predicate "a pair of numbers defines some string". Bitwise operations are naturally defined over positive integers, or, to be more precise, over the binary representations of these numbers.

Thus, nondeterministic computations with a bound on the number of steps (up to a quadratic function in the bound) are equivalent (both ways) to establishing the solvability of equations of the above types with a bound on the length of the solution.

Note that instead of logical-arithmetic equations containing the operation of bitwise multiplication of the binary representations of the arguments, it is sufficient to consider the system of equations in which the first equation is a polynomial with integer coefficients, and the remaining equations are equalities expressing the fact that the result of the bitwise multiplication of the binary representations of arguments is equal to zero. If our aim is to reduce the number of equations, then this number can be reduced to 2 in the following way.

The bitwise product of X_1 and Y_1 is equal to zero and ... and the bitwise product of X_K and Y_K is equal to zero if and only if there exists C such that each of $X_1, Y_1, \ldots, X_K, Y_K$ is less than C and $(C+1) \wedge C = 0$, and

$$(X_1 + (C+1)X_2 + \ldots + (C+1)^{K-1}X_K)$$
$$\wedge (Y_1 + (C+1)Y_2 + \ldots + (C+1)^{K-1}Y_K) = 0.$$

Finally, since an equality between a polynomial and a variable can be added to the first equation in a more or less traditional manner, both equations containing bitwise multiplication can replaced with the following two equations having a simpler form:

$$(C+1) \wedge C = 0, \quad X \wedge Y = 0,$$

where C, X, Y are positive integer variables. The binary representation of the solution of the first equation either is zero or consists only of ones. The last two equations can be regarded as an additional restriction on the quantifier prefix consisting of the existential quantifiers bounded by a quadratic function in the number of steps of the represented nondeterministic computation.

The operation of bitwise multiplication can be replaced by the divisibility of binomial coefficients. To be more precise, the following theorem holds.

Suppose that there are brackets [] in the expression Φ written in a single string. Then instead of Φ we write the formulas obtained from Φ by excluding (1) brackets only, and (2) brackets together with expressions inside them. If there are several pairs of brackets, the exclusions can be done independently. We shall say that the list of variables does not exceed some expression if each variable from the list does not exceed this expression.

THEOREM. *If for some positive integers C, C''' the predicate H is computable on a nondeterministic one-side one-tape Turing machine with a restriction on the number of steps defined by a polynomial in the input length of the form $C'''|X|^C$ for some C''', then the predicate can be represented in each of the following forms*:

$$\exists y[(\le 2^{C''}|X|^{2C})](P + P'' = 0),$$

$$\exists X_1 X_2[(\le 2^{C''}|X|^{2C})](P + P'''' + (2X_1 - 2^{X_2})^2 = 0),$$

$$\exists Y X_1 X_2[(\le 2^{C''}|X|^{2C})]\left(2 \mid 2^{P^2} + (2X_1 - 2^{X_2})^2 \cdot \binom{P_1}{P_2} + 1\right),$$

where Y is the list of variables, P is a polynomial with integer coefficients in the variables X, Y, P'' is the sum of two bitwise products of polynomials with positive integer coefficients in variables Y, P'''' is a bitwise product of polynomials with positive integer coefficients in variables X_1, X_2, Y, and P_1, P_2 are polynomials in X and Y, the symbol $\mid$ means "is a divisor of", and $\binom{P_1}{P_2}$ is the binomial coefficient.

COROLLARY. *The truth of each formula from the statement of the theorem with excluded brackets cannot be established on a one-side one-tape nondeterministic Turing machine in a number of steps bounded from above by the polynomial $C'' \cdot |X|^{C-1}$ for some C''. But the truth of these formulas can be established on a nondeterministic Turing machine in a number of steps bounded by the polynomial $C_1|X|^{4C}$ for some C_1.*

The corollary makes it possible to provide additional examples of universal search problems with polynomial lower bounds on the run time of their computation by nondeterministic Turing machines. These examples were first published in [2].

The proof of the theorem is based on a theorem from [2] and equivalencies making it possible to reduce the number of bitwise multiplications in the representation. The equality to zero of the product of two arguments is replaced by the predicate of the evenness of binomial coefficients, as in [10].

The last theorem makes more precise and, in particular, stronger a theorem of L. Adleman and K. Manders on the exact normal form for nondeterministic Turing machines (Theorem 4.4 in [11]). Those authors obtained a representation similar to the first formula written in a single string in the last theorem and, correspondingly, in the corollary to this theorem, but for the class of all predicates computable in a polynomial number of steps on nondeterministic Turing machines.

Thus, we have established polynomial bounds for establishing solvability of equations in number-theoretic form. These bounds are established even for the class of equations whose solvability in bounded numbers is a universal search problem.

In conclusion we may note that the last theorem and its corollary were presented at the 2nd All-Union seminar "Complexity Bounds for Computations", which took place in Grodno in May, 1983.

References

1. R. G. Nigmatullin, *Complexity of Boolean function*, Kazan. Gos. Univ., Kazan, 1983. (Russian)

2. N. K. Kossovskiĭ, *Polynomial bounds of the effectiveness of search for solutions of logical-arithmetic equations*, Logic and Foundations of Mathematics, Eighth All-Union Conf. "Logic and Methodology of Science", Abstracts of Reports, Vilnius, 1982, pp. 36–39. (Russian)
3. O. B. Lupanov, *On asymptotical complexity bounds for control systems*, Internat. Congr. Math. (Nice, 1970): Reports of Soviet Mathematicians, "Nauka", Moscow, 1972, pp. 162–167. (Russian)
4. S. V. Yablonskiĭ, *A survey of some results in the area of discrete mathematics*, All-Union Conf. Problems of Theoretical Cybernetics, vol. 5 (42), Scientific Council for Cybernetics, Moscow, 1970, pp. 5–15. (Russian)
5. L. A. Sholomov, *On a sequence of complexly realizable functions*, Mat. Zametki **17** (1975), 957–966; English transl in Math. Notes **17** (1975).
6. R. I. Freidzon, *Families of recursive predicates of measure zero*, Zap. Nauchn. Sem. Leningrad. Otdel. Mat. Inst. Steklov. (LOMI) **32** (1972), 121–128; English transl. in J. Soviet Math. **6** (1976), no. 4.
7. V. L. Temov and M. B. Matskin, *Representation of some search problems in the technique of Boolean matrices*, Third Conf "Application of the Methods of Mathematical Logic", Abstracts of Reports, Tallinn, 1983, pp. 159–160. (Russian)
8. O. B. Lupanov, *On the synthesis of some classes of control systems*, Problemy Kibernetiki **10** (1963), 88–96. (Russian)
9. N. K. Kossovskiĭ, *Elements of mathematical logic and its application to the theory of recursive algorithms*, Izdat. Leningrad. Univ., Leningrad, 1981. (Russian)
10. Yu. V. Matiyasevich, *A new proof of the theorem about the exponentially Diophantine representation of listable predicates*, Zap. Nauchn. Sem. Leningrad. Otdel. Mat. Inst. Steklov. (LOMI) **60** (1976), 75–92; English transl. in J. Soviet Math. **14** (1980), no. 5.
11. L. Adleman and K. Manders, *Diophantine complexity*, Proc. 17th Annual IEEE Sympos. Foundations of Computer Science, New York, 1976, pp. 81–88.

On a Class of Polynomial Systems of Equations Following from the Formula for Total Probability and Possibilities for Eliminating Search in Solving Them

An. A. Muchnik

ABSTRACT. A generalization of linear systems of equations for stable states of discrete random processes is studied. The effectiveness of two classical methods of approximating a solution is examined.

1. Introduction

This paper is directed towards those readers who are meeting in practice with the problem considered below. The purpose is to give the most complete and independent exposition possible, requiring only minimal prior knowledge.

The mathematical model we shall be investigating was proposed by Petrovskiĭ. The result of Section 2 belongs to Vysotskiĭ. It is based on Brouwer's theorem, whose proof relies on Sperner's formula. The main arguments in Section 4 are similar to the approximation of continuous functions by polynomials using Bernstein's method. In Sections 5 and 6 we study conditions determining the convergence rate for the methods of iterations and descent. The computation of the number of ladders in Section 8 is an exercise in combinatorics.

Various information related to this subject can be found in [1].

New results contained in this article were presented at a seminar in the Center on the Problems of Development and Control in Humanities of the Ministry of Culture in 1982.

The author is grateful to S. F. Soprunov for stimulating discussions.

1. Statement of the problem

1.1. Preliminary remarks. Let P be a finite collection of finite sets; the sets of the collections will be called problems, and their elements will be called states. With each state we associate a variable which will be interpreted as the probability of the event that the problem containing this state lies precisely in this state. Our purpose is to find the probabilities of all states.

©1996 American Mathematical Society

The algebra of events of the probability space is given by the collections of states, one from each problem. The event corresponding to a state consists of those collections that select this state from the corresponding problem.

Denote the event corresponding to the state S by $m(S)$ and the probability of the event μ by $w(\mu)$.

Let S be a state, $U \subset P$. Then it follows from a well-known theorem that

$$w(m(S)) = \sum_{v \in \otimes U} w(m(S) \mid \bigcap_{p \in U} m(\pi_p v)) \times (\bigcap_{p \in U} m(\pi_p v)),$$

where $\otimes U$ is the direct product of sets belonging to U and $\pi_p v$ is the projection of U on p. If we know the way in which problems in U influence the problem containing S, then we can assume that we are given conditional probabilities

$$w(m(s) \mid \bigcap_{p \in U} m(\pi_p v)).$$

If the problems in U are independent, then

$$w(\bigcap_{p \in U} m(\pi_p v)) = \prod_{p \in U} w(m(\pi_p v)).$$

Thus the two preceding conditions yield equations on the state probabilities. The verification of whether these conditions are satisfied must be carried out in each case arising in practice.

Formal description. Let P be a finite set of finite sets (problems) whose elements are variables with the domain $[0,1]$. In addition, let $\sum_{x \in p} X = 1$ for each $p \in P$. Suppose we are given a function U associating with each problem a set of problems (influencing it). Suppose that a function is given associating with each variable X ($\in p \in P$) and each collection $v \in \otimes U(p)$ the number $W_x(v) \in [0,1]$. Here $\sum_{x \in p} W_x(v) = 1$ for all $p \in P$ and all $v \in \otimes U(p)$. We must solve the following system of equations:

(1) $$\left\{ x = \sum_{v \in \otimes U(p)} W_x(v) \prod_{q \in U(p)} \pi_q(v) \mid x \in p \in P \right\}.$$

2. Existence of a solution

The system of equations (1) can be written in general form as $y = f(y)$, where y is a vector and f is an operator. Let us investigate the properties of f.

Domain. As we know, the variables of a problem take values in a simplex whose dimension is one less than the number of variables in the problem. Therefore, the domain of definition S for the operator f (the domain of definition of all variables) is the direct product of simplices.

PROPOSITION. *The simplex is homeomorphic to a cube of the same dimension.*

PROOF. Let n be the dimension. Embed the simplex and the cube in R^n so that the centers of both are at the origin $\vec{0}$. Denote the resulting figures by σ and κ respectively. Consider the functions $\alpha, \beta \colon R^n \setminus \{\vec{0}\} \to (0, \infty)$:

$$\alpha(y) = \sup_{ky \in \sigma} k, \quad \beta(y) = \sup_{ky \in \kappa} k.$$

Evidently, these functions are continuous and the ratios $\frac{\alpha}{\beta}, \frac{\beta}{\alpha}$ are bounded. The mapping G defined by

$$G(y) = \begin{cases} \frac{\beta(y)}{\alpha(y)} y & \text{for } y \neq \vec{0}, \\ \vec{0} & \text{for } y = \vec{0} \end{cases}$$

is a homeomorphism of σ on κ.

Completion of the proof. The direct product of cubes is a cube whose dimension is equal to the sum of the dimensions of the factors. This and the proof of the statement imply that S is homeomorphic to a simplex of a certain dimension.

Range. Let us denote the xth coordinate of the vector $f(y)$ by $fx(y)$,

$$fx(y) = \sum_{v \in \otimes U(p)} W_x(v) \prod_{q \in U(p)} \pi_q(v),$$

where $x \in p \in P$. We have $fx(y) \geq 0$ for all y and for all x, since $fx(y)$ it is obtained from positive numbers by the operations of multiplication and addition. Also we have $\sum_{x \in p} fx(y) = 1$ for all y and all p. Indeed,

$$\sum_{x \in p} fx(y) = \sum_{x \in p} \left(\sum_{v \in \otimes U(p)} W_x(v) \prod_{q \in U(p)} \pi_q(v) \right)$$

$$= \sum_{v \in \otimes U(p)} \left(\sum_{x \in p} W_x(v) \prod_{q \in U(p)} \pi_q(v) \right)$$

$$= \sum_{v \in \otimes U(p)} \left(\left(\sum_{x \in p} W_x(v) \right) \prod_{q \in U(p)} \pi_q(v) \right)$$

$$= \sum_{v \in \otimes U(p)} \prod_{q \in U(p)} \pi_q(v) = \prod_{q \in U(p)} \sum_{z \in q} z = \prod_{q \in U(p)} 1 = 1.$$

Therefore, the operator f maps the domain S into itself.

Continuity of the operator. The operator f is continuous since it is defined in terms of operations of multiplication and addition.

If ψ is a homeomorphism of the domain S onto the simplex, then finding a fixed point for the operator f is equivalent to finding a fixed point of the continuous operator $\psi^{-1} f \psi$ mapping the simplex onto itself. Let us first state the problem in discrete language. This is more than appropriate if we are looking for a computer-generated solution.

Partition of the simplex. We shall identify an n-simplex with the set of its vertices $(a_0, a_1, \ldots, a_{n-1}, a_n)$.

With each point b in the simplex $(a_0, a_1, \ldots, a_{n-1}, a_n)$ we associate *its proper partition* of the simplex into the simplices $(b, a_1, \ldots, a_{n-1}, a_n)$, $(a_0, b, \ldots, a_{n-1}, a_n)$, $\ldots$, $(a_0, a_1, \ldots, b, a_n)$, $(a_0, a_1, \ldots, a_{n-1}, b)$.

A face of a simplex is a simplex of dimension one less whose vertices are also vertices of the first simplex. By a partition we mean a partition of the simplex into simplices of the same dimension.

Given a partition of a face, it can be extended to the partition of the simplex in such a way that the vertices of each element in the partition are the vertices of some element of the partition of the face and a vertex of the simplex that does not belong to this face.

If a point is fixed in the interior of the simplex, and for each face a partition is selected, then we can extend a partition of each face to a partition of the corresponding element in the fixed point's proper partition. This results in a partition of the simplex that extends both its proper partition and given partitions of the faces.

If a point is fixed in the interior of each simplex of positive dimension, then we can fix a partition for each simplex by induction in the dimension. For the simplex of dimension 0 there is a unique partition. The induction step is performed according to the above procedure.

It is easy to prove by induction that each face of an element of the fixed partition either lies on the face of the simplex or is a face of another element of the partition. If each element of the partition is partitioned in some manner, this results in the partition of the initial simplex such that the *two-sided property* given in the preceding sentence is preserved. This partition process can be continued indefinitely. We want the size of the elements of the partition to tend to zero. In order to achieve this goal, fix the arithmetic mean of the vertices in the simplex $\frac{a_0 + a_1 + \ldots + a_{n-1} + a_n}{n+1}$. Let us estimate the distance from this center point to the vertices of the simplex in terms of the difference between vertices of the simplex. For example, the distance to a_0 is

$$\left| \frac{a_0 + a_1 + \ldots + a_{n-1} + a_n}{n+1} - a_0 \right|$$
$$= \left| \frac{(a_1 - a_0) + (a_2 - a_0) + \ldots + (a_{n-1} - a_0) + (a_n - a_0)}{n+1} \right|$$
$$\leq \frac{n \max |a_i - a_0|}{n+1} \leq \left(1 - \frac{1}{n+1}\right) \max_i |a_i - a_0|.$$

Let us call the maximum of the distances between the vertices of the simplex its diameter. Thus, the distance from the center to a vertex (and, consequently, to any point of the simplex) does not exceed $\left(1 - \frac{1}{n+1}\right)$ times the diameter of the simplex. Hence it follows that the diameter of the elements of the fixed partition is bounded by the same number. So by partitioning the simplex in a fixed manner k times we obtain a partition with all elements having diameter not exceeding $\left(1 - \frac{1}{n+1}\right)^k$ times the diameter of the initial simplex.

Let φ be a continuous operator mapping the simplex into itself. With each point in the simplex we can associate the nondegenerate element of its proper partition which includes its image (an element can be degenerate if a point belongs to a face).

If there are several such elements, select one of them. An element of the proper partition is uniquely defined by that face of the partitioned simplex on which it is based. A face of the simplex is uniquely defined by the vertex that does not belong to it. Therefore, with each point of the simplex we have associated its vertex. This correspondence satisfies the following *boundary condition*: points lying on a face can only correspond to vertices belonging to the same face. It turns out that even this very "rough" function is sufficient to find the fixed point of the initial function (of the operator φ). If $\varphi(y) \neq y$, then the ray $\overrightarrow{y, \varphi(y)}$ intersects the face of the simplex corresponding to the point y. The intersection of all faces of the simplex is empty. Indeed, the simplex is defined by the equation $\sum_{i=0}^{u} x_i = 1$, where $x_i \geq 0$. A face is defined by the equation $x_i = 0$. Thus, if $\varphi(y) \neq y$, then the ray $\overrightarrow{y, \varphi(y)}$ does not intersect some face U. Since φ is continuous, we have $\varphi(Z) \neq Z$ for all Z sufficiently close to y, and the ray $\overrightarrow{Z, \varphi(Z)}$ does not intersect the same face. Thus, if arbitrarily close to point y there are points corresponding to all possible vertices, then $\varphi(y) = y$.

PROPOSITION. *Suppose there is a simplex and a function which associates to the vertices of the elements of the partition the vertices of the partitioned simplex. Suppose also that the two-sided property is satisfied for the partition, and the boundary property for the function.*

Then there exists an element in the partition in whose vertices the function positions all the vertices of the partitioned simplex.

PROOF. We shall prove by induction on the dimension that there exist an *odd number* of desired elements. The 0-simplex has exactly one desired element. To perform the step from n to $n+1$, let us assume that the vertices of the partitioned simplex are $a_0, a_1, \ldots, a_n, a_{n+1}$. A face in the element of the partition will be called regular, if the function positions $a_0, a_1, \ldots, a_n$ in its vertices. By the boundary condition regular faces occur only on the face $a_0, a_1, \ldots, a_n$ and inside the simplex. It is easy to check that the boundary condition implies the boundary condition for each of the faces of the partitioned simplex. The two-sided property is also satisfied for faces. Therefore, the inductive hypothesis implies that the number of regular faces on the face $a_0, a_1, \ldots, a_n$ is odd. Consider an element of the partition having a regular face. If the vertex of the element which does not belong to this face is mapped into a_{n+1}, then the element is the desired one, and there is one regular face in it. Otherwise the element is not the desired one and has two regular faces. Let us sum up all the regular faces over all elements of the partition. By the two-sided property, all internal faces will be counted twice, while the number of regular faces on the boundary, as was already proved, is odd; therefore the entire sum is odd. Hence the number of the desired element of the partition is odd.

If the diameter of the elements of the partition tends to 0, then (since the simplex is compact) there exists a limit point for the sequence of contracting desired elements. This point is a fixed point for the operator φ.

3. Existence of the algorithm

Given a partition and a function, one can use a search process to find the desired element of the partition. As a rule, we can determine how small a partition should be in order for the operator φ to move the points of the desired element of the partition by less than ε. Given $\varepsilon > 0$, we must find $\delta > 0$ such that for any $|y_1 - y_1| < \delta$ the relation $|\varphi(y_2) - \varphi(y_1)| < \varepsilon$ is satisfied. Such a δ exists since any continuous operator on a compact is uniformly continuous. On the other hand, it is usually impossible to specify how small a partition should be in order that the distance between the desired element of the partition and the fixed point be less than ε (the work of V. P. Orevkov [**2**]).

4. Problem complexity

We will show that our main problem is as hard as the general problem of finding the fixed point of a continuous mapping of a simplex into itself.

Relation between probability and relative frequency. Suppose we have a random variable assuming values 0 and 1. Perform n independent trials of the variable. One can prove that the probability that the relative frequency of the occurrence of 1 and the probability of the occurrence of 1 differ by more than ε, is less than $1/c \cdot \varepsilon e^{2\varepsilon^2 n}$, where c can be taken just above $4\sqrt{2\pi}$ ("just above" $\to 0$ as $n \to \infty$). This estimate tends to 0 as $n \to \infty$ and does not include the probabilities of values for the random variable. Now let ψ be a random variable assuming k values $a_1, \ldots, a_k$. For each i we can consider the random variable taking value 1 if $\psi = a_i$ and 0 otherwise. Applying the above estimate to these variables, we see that the probability that the relative frequency of the event that at least one value of ψ in n independent trials differs from the probability of the same value by no more than ε, does not exceed $k/c \cdot \varepsilon \cdot e^{2\varepsilon^2 n}$.

Let us proceed with the construction of system (1).

Construction. The set P will consist of the problem A and the problems $B_1, \ldots, B_n$. Each of the problems will have k variables. The variables of the problem B_j will be denoted by $(b_{j_1}, \ldots, b_{j_k})$. The vector of the variables for a problem will be denoted in the same way as the problem itself.

The variables of the problem A are interpreted as the probabilities of values of the variable Ψ. Suppose that a sequence of n independent trials of ψ is given. Then b_{ij} is the probability of the event that $\psi = a_i$ in trial j.

Let us proceed with the formal construction. Each B-problem is influenced by one problem A, while A is influenced by all B_j.

The comparisons for B-problems are of the form $B_j = A$. The interpretation of this is that the same random variable is realized in all trials.

In vector notation, equations for the problem A are of the form:

$$A = \sum_{v \in \otimes w(v)} \prod_{j=1}^{n} \pi_j(v),$$

where $\otimes B$ is the direct product of $B_1, \ldots, B_n$.

Substituting in the right-hand side for the variables of the B-problem the corresponding variables of the problem A, we obtain a fixed point problem for a $(k-1)$-simplex.

Let f be a continuous mapping of the $(k-1)$-simplex into itself. An appropriate choice of the coefficients $W(v)$ makes the operator $\sum_{v \in \otimes B} w(v) \prod_{j=1}^{n} \pi_j(v)$, where for the variables of the B-problems are substituted the corresponding variables of problem A, uniformly close to f. The larger n, the better the approximation.

To each element $v \in \otimes B$ there corresponds a sequence of values of the variable ψ. Denote the vector of frequencies of occurrences of $a_1, \ldots, a_k$ by $h(v)$. Evidently, $h(v)$ lies in the domain of definition of f. Set $w(v) = f(h(v))$. Define the norm of the vector as the maximum of absolute values of its coordinates. A continuous mapping of a compact is uniformly continuous. Therefore, there exists ε such that for any points y_1, y_2 of the simplex the relation $|y_1 - y_2| \leq \varepsilon$ implies

$$|f(y_1) - f(y_2)| \leq \delta.$$

Note that $\prod_{j=1}^n \pi_j(v)$ is the probability that the sequence of values ψ obtained in n independent trials coincides with v.

Substitute for the value of the variables in the sum

$$\sum_{v \in \otimes B} f(h(v)) \prod_{j=1}^n \pi_j(v)$$

a point of the simplex y. Then the sum splits into two parts: over such v for which $|h(v) - y| > \varepsilon$, and over such v for which $|h(v) - y| \leq \varepsilon$. The first part of the sum is small because

$$\sum_{\substack{v \in \otimes B \\ |h(v)-y|>\varepsilon}} \prod_{j=1}^n \pi_j(v)$$

after the substitution for b_{j_i}, expresses the probability of the event that the vector of frequencies differs from the vector of probabilities by more than ε (for at least one value of ψ). The coefficients of $f(h(v))$ are bounded. In the second part of the sum we have $|f(h(v)) - f(y)| \leq \delta$ (uniform continuity). The expression

$$\sum_{\substack{v \in \otimes B \\ |h(v)-y|\leq\varepsilon}} \prod_{j=1}^n \pi_j(v)$$

after the substitution of $\pi_i(y)$ for b_{j_i} gives the probability of the event that the frequencies are close to the probability of y, which is thus close to 1. Hence the entire sum is close to $f(y)$, as required.

Reduction to the quadratic case. In the preceding construction the degree of equations was equal to n (the number of trials) and increased with the growth in the accuracy of the approximation of the mapping f. The number of variables in the problem was the same number k. It turns out that the construction can be modified in such a way that the degree of equations is equal to 1 or 2, although at the expense of having more variables.

Namely, starting with system (1) we can construct an equivalent system of the same type in which each equation has degree 1 or 2. For each problem p, introduce a new problem p' whose variables are indexed by elements of $\otimes U(p)$. To each equation of the old system there corresponds a linear equation of the new system

$$x = \sum_{v \in p'} W_x(v) \cdot v, \quad x \in p \in P.$$

It remains to add the condition that $v = \prod_{q \in U(p)} \pi_q(v)$ for $v \in p'$. If A and B are problems, then we can construct a problem whose variables are denoted by the element $A \emptyset B$, with the equations $\langle a, b \rangle = a \cdot b$, $b \in B$. The degree of these equations is 2. By applying these equalities (i.e., introducing new problems with corresponding equations) as many times as there are problems in $U(p)$ (to be more precise, one time less), we obtain the required equalities for $v \in p'$.

5. Iterative method

If μ is a metric space and f a continuous mapping of μ into itself, then the iterative method for finding a fixed point of f is an attempt to find the limit of $f^n(x_0)$ as $n \to \infty$, where $x_0 \in \mu$. If this limit exists, then it is the fixed point of f. If f is a contraction mapping with coefficient α (i.e., $0 \le \alpha < 1$ and $\rho(f(x), f(y)) \le \alpha \rho(x, y)$ for all $x, y \in \mu$, where ρ is the distance in μ), then $\rho(f^{n+1}(x_0), f^n(x_0)) \le \alpha^n (\rho(f(x_0), x_0)$ for $n \ge 0$. Therefore,

$$\rho(f^{n+k}(x_0), f^n(x_0)) \le \alpha^n \rho(f(x_0), x_0) \cdot \sum_{i=0}^{k-1} \alpha^i$$

$$= \alpha^n \rho(f(x_0), x_0) \cdot \frac{1-\alpha^k}{1-\alpha} \le \frac{\alpha^n}{1-\alpha} \rho(f(x_0), x_0)$$

for all $n \ge 0$, $k > 0$. Thus, $\rho(f^{n+k}(x_0), f^n(x_0))$ tends to 0 as $n \to \infty$, and the smaller α is, the faster. If μ is a complete space, then the limit of $f^n(x_0)$ as $n \to \infty$ exists and $\rho(f^n(x_0), x) \le \frac{\alpha^n}{1-\alpha} \rho(f(x_0), x_0)$.

Differentiable case. In our case μ is the product of simplices (which is a complete space), f a polynomial (and, consequently, differentiable) mapping.

In general, if f is a continuous differential mapping from a normed vector space A into a normed vector space B, then f is a contraction map with the coefficient α if and only if the derivative of f at each point is a contracting linear operator with the coefficient α. The "only if" part is proved by differentiating f, the "if" part by integrating f'.

In our case the derivative acts in the linear space D tangent to the domain of definition of f (i.e., to S).

Let us represent vectors in D in the basis of initial variables. Denote the projection of the vector y on the direction corresponding to the variable x by y_x.

$$D = \left\{ y \mid \forall p \in P \sum_{x \in p} y_x = 0 \right\}.$$

Norm. Introduce a norm on vectors $\|y\| = \max_{p \in P} \sum_{x \in p} |y_x|$. (Prove that it is a norm!) Our aim is to estimate the norm of f' (uniformly over the points at which the derivative is taken) with respect to the norm on vectors. If the norm of f' does not exceed α (< 1) at all points, then f is a contracting mapping with coefficient α.

Bound for the norm of f'. Consider a point x in S, Its coordinates are values of variables X. Let us estimate the norm of the derivative of f at this point applied to the vector y ($\in D$) in terms of the norm of y. Since $\|f'(y)\| =$

$\max_{p \in P} \sum_{x \in p} |f'_x(y)|$, it is sufficient (and necessary) to estimate $\sum_{x \in p} |f'_x(y)|$ for a single p.

This is what we shall do now.

By definition
$$f_x = \sum_{v \in \otimes U(p)} W_x(v) \prod_{q \in U(p)} \pi_q(v).$$

An application of the chain rule formula yields
$$f'_x(y) = \sum_{v \in \otimes U(p)} W_x(v) \sum_{r \in U(p)} y_{\pi_r}(v) \prod_{\substack{q \in U(p) \\ q \neq r}} \pi_q(v).$$

Using the fact that $\sum_{z \in r} z = 1$, we can multiply each $y_{\pi_r}(v)$ by $\sum_{z \in r}$. Thus,
$$f'_x(y) = \sum_{v \in \otimes U(p)} W_x(v) \sum_{r \in U(p)} y_{\pi_r}(v) \left(\sum_{z \in r} z \right) \prod_{\substack{q \in U(p) \\ q \neq r}} \pi_q(v).$$

Making use of the distributive property in the right-hand side, we obtain a linear combinations of monomials $\prod_{q \in U(p)} \pi_q(v)$, where $V \in \otimes U(p)$. Namely,
$$f'_x(y) = \sum_{v \in \otimes U(p)} \left(\sum_{q \in U(p)} \sum_{z \in q} w_x(v_q^z) y_z \right) \prod_{q \in u(p)} \pi_q(v),$$

where v_q^z denotes the result of substituting z for the qth coordinate of v. Since the variables x are nonnegative, we have
$$|f'_x(y)| \leq \sum_{v \in \otimes U(p)} \left(\sum_{q \in U(p)} \left| \sum_{z \in q} w_x(v_q^z) y_z \right| \right) \prod_{q \in u(p)} \pi_q(v).$$

Define $m_x(v_q) = \min_z w_x(v_q^z)$. Since $y \in D$, we have $\sum_{z \in q} y_z = 0$, which implies that
$$\left| \sum_{z \in q} w_x(v_q^z) y_z \right| = \left| \sum_{z \in q} (w_x(v_q^z) - m_x(v_q)) y_z \right|.$$

Since the values $(w_x(v_q^z) - m_x(v_q))$ are nonnegative, we have
$$|f'_x(y)| \leq \sum_{v \in \otimes U(p)} \left(\sum_{q \in U(p)} \left(\sum_{z \in q} w_x(v_q^z) |y_z| - m_x(v_q) \sum_{z \in q} |y_z| \right) \right) \prod_{q \in U(p)} \pi_q(v).$$

Making use of the last inequality to estimate $\sum_{x \in p} |f'_x(y)|$, move the sign of summation over X inside. This gives us
$$\sum_{x \in p} |f'_x(y)| \leq \sum_{v \in \otimes U(p)} \left(\sum_{q \in U(p)} \left(\left(1 - \sum_{x \in p} m_x(v_q) \right) \sum_{z \in q} |y_z| \right) \right) \prod_{q \in U(p)} \pi_q(v).$$

Define $\nu(p)$ as the cardinal number of the set $U(p)$, and
$$\mu(p) = \min_{v \in \otimes U(p)} \sum_{q \in U(p)} \sum_{x \in p} m_x(v_q).$$

Since $\sum_{z\in q}|z_q| \leq \|y\|$, we have

$$\sum_{x\in p}|f'_x(y)| \leq \sum_{v\in \otimes U(p)} \prod_{q\in U(p)} \pi_q(V)$$
$$= (\nu(p) - \mu(p))\|y\| \sum_{v\in \otimes U(p)} \prod_{q\in U(p)} \pi_q(v) = (\nu(p) - \mu(p))\|y\|.$$

Thus, $\|f'\| \leq \max_{p\in P}(\nu(p) - \mu(p))$.

6. Method of descent

Suppose that $\varphi: S \to R$ is a differentiable functional and that our task is to find its minimal value. In order to apply the known descent methods, it is useful to estimate $\min_{y\in S}(\varphi(z))'(y-z)$ in terms of $\varphi(z)$, where $z \in S$ and $(\varphi(z))'$ is the derivative of φ at the point z. The idea of the descent method is that a translation along the vector $(y-z)$ reduces the value of φ.

In our case $\varphi(z) = (f(z) - z, f(z) - z)$, where $(\cdot, \cdot)$ is the usual scalar product. Clearly, $\varphi(z) \geq 0$, and $\varphi(z) = 0 \leftrightarrow f(z) = z$. We have

$$(\varphi(z))'(y-z) = 2(f(z) - z, f(z) - z)'(y-z))$$
$$= 2(f(z) - z, (f(x))'(y-z) + f(z) - y)$$
$$= -2(f(z) - z, f(z) - z) + 2(f(z) - z, (f(z))'(y-z) + f(z) - y)$$
$$= -2\varphi(z) + (f(z) - z, (f(z))'(y-z) + f(z) - y).$$

Now our goal is to find $y \in S$ such that

$$(f(z))'(y-z) + f(z) - y = 0.$$

For this value of y we have $(\varphi(z))'(y-z) = -2\varphi(z)$. We want to prove that there exists $y \in S$ such that $(f(z))'(y-z) + f(z) = y$, i.e., the fixed point of the affine (with respect to y) operator $(f(z))'(y-z) + f(z)$ in the domain S. Since an affine operator is continuous, it is sufficient to prove that it maps the domain S into itself. Note that since $y \in S, z \in S$, we have $(y-z) \in D$ and, consequently, also $(f(z))(y-z) \in D$. Since $f(z) \in S$, for each $p \in P$ we have

$$\sum_{x\in p}((f(z))'(y-z) + f(z))_x = 1.$$

Therefore, it remains to prove that

$$((f(z))'(y-z) + f(z) + f(z))_x \geq 0$$

for each $x \in p \in P$. Of course, this inequality does not always hold. We find a simple necessary and sufficient condition for it to be true.

Let $p \in P$ and $x \in p$. We need

$$(f(z))'_x(y) \geq (f(z) - f_x(z))$$

for all $y, z \in S$. Since f_x is a homogeneous polynomial of degree $\nu(p)$, we have $(f(z))'_x(z) = \nu(p)f_x(z)$. This can also be easily established by a direct application

of the derivation formula. In the process of estimating the norm of f', we proved that
$$(f(z))'_x(y) = \sum_{v \in \otimes U(p)} \left(\sum_{q \in U(p)} \sum_{t \in q} W_x(v_q^t) y_t \right) \prod_{q \in U(p)} \pi_q(v).$$
Thus, we need that the equality
$$\sum_{v \in \otimes U(p)} \left(\sum_{q \in U(p)} \sum_{t \in q} W_x(vt_q) y_t \right) \prod_{q \in U(p)} \pi_q(v) \geq (\nu(p) - 1) f_x(z)$$
holds for all $y, z \in S$. Since
$$f_x(z) = \sum_{v \in \otimes U(p)} w_x(v) \prod_{q \in U(p)} \pi_q(v),$$
we have an inequality for linear combinations of $\prod_{q \in U(p)} \pi_q(v)$. Substituting for each $v \in \otimes U(p)$ that value of z for which $\prod_{q \in U(p)} \pi_q(v) = 1$ and the remaining products are equal to 0, we see that the inequality must be true coefficientwise (evidently, it is also sufficient). Thus, for all $y \in S$ and all $v \in \otimes U(p)$ we have
$$\sum_{q \in U(p)} \sum_{t \in q} W_x(v_q^t) y_t \geq (\nu(p) - 1) w_x(v).$$
Substituting those y for which in every q there exists t such that $y_t = 1$ and the remaining coordinates are equal to 0, we see that
$$\sum_{q \in U(p)} \min_{t \in q} w_x(v_q^t) \geq (\nu(p) - 1) W_x(v).$$
Clearly, this relation is also sufficient. Thus, a necessary and sufficient condition is
$$\sum_{q \in U(p)} m_x(v_q) \geq (\nu(p) - 1) W_x(v).$$

Norm of f'. If we sum up the latter inequalities over all $x \in p$, we obtain $\sum_{x \in p} \sum_{q \in U(p)} m_x(v_q) \geq \nu(p) - 1$, and $\mu(p) - \nu(p) - 1$. In this case $\|f'\| \leq 1$. This (and the next example) illustrates the comparative efficiency of the iterative and the descent methods.

Example. Let $P = \{\{x_1, x_2\}\}$. Consider the system
$$\left\{ \begin{array}{c} x_1 = x_2 \\ x_2 = x_1 \end{array} \right\}.$$
It is equivalent to the equation $x = 1 - x$. The iterative method does not work in this case, but the method of descent gives fast convergence to the solution $x = 0.5$.

Finding $m_x(v_q)$. When computing $m_x(v_q)$ we have to bear in mind that $m_x(v_q)$ does not depend on the qth coordinate of v. Therefore, one can organize the table of "truncated" vectors $\bar{v}$ with one coordinate dropped. First we perform a search process in the lexicographic order among the vectors without the first coordinate, then without the second, and so on. Each vector v is "connected" with $\nu(p)$ "truncated" vectors. Performing the search in the lexicographic order, we shall replace every number located in the squares of the table connected with v by the minimum of its $uw_x(v)$. This procedure reduces the search.

7. Main ideas

Two main ideas were used to derive the results in this article: transformation of the formula for $f'_x(y)$ in order to find the degree of the polynomial, and implicit determination of the desired value for y in the justification of the descent method.

8. Multiple influence

Both main ideas are also applicable if we assume that one problem can influence another n times (i.e., its variables occur n times in the vector v).

When vectors v are stored in the computer memory, there is the problem of saving memory, since the order of coordinates is not always essential. Consider the following problem. Suppose that in the vector v, all copies of the variable with a smaller number precede all copies of the variable with larger number, i.e., the vector starts with the copies of the first variable, then the copies of the second variable, and so on. (The variables in the problem under consideration are assumed to be ordered in some way.) This creates the property of uniqueness for the representation. We shall call such vectors canonical.

Let the variables belonging to the problem under consideration be $x_1, \ldots, x_k$. Then the number of all sequences of these variables of length n is equal to k^n. A sequence in which the numbers of variables are nondecreasing will be called a *ladder*. Let us compute the number of all ladders. Consider n cells, in which the variables can be positioned. Distribute $k-1$ walls among the cells in such a way that the ith wall separates the group of variables x_i from the group of variables x_{i+1} (two walls can to be adjacent to each other if the corresponding group of variables in empty). Evidently, the ladders and the placements of walls are in one-to-one correspondence. Consider $(n + k - 1)$ positions such that each of them can be occupied by either a cell or a wall. By placing cells in n positions and walls in $(k-1)$ positions we obtain a placement of walls. It is now evident that the desired number of ladders is equal to

$$\binom{n}{n+k-1} = \frac{(n+k-1)!}{n!(k-1)!}$$

(the binomial coefficient).

The number of canonical vectors is evidently equal to the product of the number of ladders for all problems of influence.

It will be convenient to store ladders in the computer memory in reverse lexicographic order. Then the address of a ladder is equal to the number of ladders (it is now more convenient to speak about $(n-k)$-ladders) greater than the given one. The set of $(n-k)$-ladders that are greater that the given one is decomposed into n sets: the ith set contains the ladders that are equal to the positions with numbers less than i, and is greater than the given ladder at the position i. If the ith position in the ladder is occupied by the variable x_j, then it is not hard to see that the number of ladders in the ith set corresponding to this ladder is equal to the number of all $(n-i+1)(k-j)$-ladders. (For $\chi = 0$ the number ν of χ-ladders is equal to 0.)

If a canonical vector v contains variables of the problems from the set U, then the address of v is equal to

$$\sum_{q \in U} \alpha(v, q) \prod_{\substack{r \in U \\ r < q}} \beta(r),$$

where $\alpha(v, q)$ is the address of the ladder of variables of the problem q in the vector v, $\beta(r)$ is the number of all ladders of variables of the problem r; we assume that an ordering $r < q$ is defined on the problems; and the product of the empty set of factors is equal to 1.

For a fast computation of addresses of canonical vectors, two tables should be created in the computer memory: the table of the numbers $\binom{\kappa}{\nu}$, where $0 \leq \kappa\nu <$ max(multiplicity of the influence of q) + (the number of variables in q); and the table of numbers $\prod_{\substack{r \in U \\ r \leq q}} \beta(r)$, where $q \in U$.

References

1. V. S. Vysotskiĭ and S. A. Petrovskiĭ, *A study of some questions of existence and uniqueness of solutions for systems of equations in the nonlinear analysis model on problem nets*, Analysis on Problem Nets, vol. 1, Moscow, 1980, pp. 83–101. (Russian)
2. V. P. Orevkov, *A uniformly continuous constructive mapping of a square into itself with no fixed points*, Appendix 6 to the Russian transl. of J. Barwise (ed.), *Handbook of mathematical logic. Part IV: Proof theory and constructive mathematics*, "Nauka", Moscow, 1982. (Russian)

S. Maslov's Iterative Method: 15 Years Later (Freedom of Choice, Neural Networks, Numerical Optimization, Uncertainty Reasoning, and Chemical Computing)

V. Kreinovich

ABSTRACT. In 1981, S. Maslov proposed a new iterative method for solving propositional satisfiability problems. The 1981–87 results related to this method were described in the present book. In this chapter, we briefly recall the origins of Maslov's method, and describe further results and related ideas.

1. 1979–81: The origins of S. Maslov's iterative method

1.1. Freedom of choice is necessary.

1.1.1. *Freedom of choice is necessary in social life.* On the top of a high mountain, with hardly enough air to breathe, one fully realizes the importance of breathing. In a totalitarian state, where even elections had exactly one candidate to choose from, we realized the importance of freedom; and one of the main aspects of freedom is the freedom of choice.

Freedom may be seen as a burden:

- In science and *engineering* (e.g., when we plan a flight to a distant planet), we:
  - formulate a reasonable optimization problem (e.g., getting there faster, or minimizing the cost),
  - solve it, and then
  - follow the optimal trajectory.

  In this case, the "freedom" will mean the freedom to deviate from this optimal trajectory. Such a deviation will only make things worse and can even lead to a disaster.

- In *economics*, at first glance, all we have to do is:
  - translate the ideas of universal happiness and prosperity into a reasonable objective functions,
  - solve the corresponding optimization problem, and then
  - follow the optimal economic plan.

©1996 American Mathematical Society

In this case, the freedom of choice can only mean the freedom to deviate from this optimal plan, and thus, to make life worse. Such freedom leads to chaotic perturbations that have often been used to explain economic crises (including the Great Depression).

This argument may sound good in theory, but practical experience shows that the communist economic system, based on this idea, has failed, while the capitalist economies, based on the built-in freedom of choice, have been doing incomparably better. So, *freedom of choice is needed*.

1.1.2. *Freedom of choice is compatible with modern physics.* Even in physics (from which the idea of determinism originated), modern theories are no longer formulated in terms of differential equations (equations that uniquely describe the future). Starting from the quarks, physical theories are mainly formulated in terms of symmetry groups and other indirect characteristics that, usually, do not lead to a unique prediction of the future.

According to modern physics, the evolution of the Universe is not uniquely predetermined by its current state. In this sense, for us, *there is freedom of choice*.

1.1.3. *Freedom of choice is necessary in computer science.* In the *social sciences* and in *modern physics*, the necessity of freedom is a natural and reasonably old idea. In contrast, in *computer science*, the idea of "freedom of choice" is reasonably new.

Traditionally, the main goal of computer science was to *automate*: to automate the control of a complicated plant, to automate the solution of complicated problems, etc. If our goal is to automate, then, of course, we do not want any human involvement in the resulting process. This process must be *algorithmic*, with all the steps uniquely predetermined by the initial program and by the input data. In an algorithm, there is no room for freedom.

If for a given problem we can find an algorithm, and not only *an* algorithm, but a *feasible* algorithm that takes a reasonable time to run, then the problem is solved. It turns out, however, that for many problems, a general feasible algorithm seems to be impossible. Such problems (called *intractable*, or *NP-hard*) were discovered independently by Cook, Carp, and Levin (for the first publication, see [**4**]; for a survey of such problems, see [**19**]). Each of these "intractable" problems has the following property: if we can solve this problem in reasonable (polynomial) time, then we would be able to design an algorithm that solves a huge class of discrete problems in polynomial time. This class is called NP, and hence, intractable problems are also called *NP-hard* problems.

The general belief is that a general polynomial time algorithm for solving all the problems from the class NP (a *universal problem solver*) is impossible. If this general algorithm is really impossible, this means that NP-hard problems cannot be solved in polynomial time.

Historically the first problem for which intractability (in this sense) has been proven is the *propositional satisfiability problem*:

- given a *propositional formula*, i.e., a Boolean ("and", "or", "not") combination of Boolean ("true-false") variables),
- to find the values of these variables (if any) for which the given formula becomes true.

For intractable problems, there is no *algorithm* that will solve all particular cases of such problems in reasonable time. So, our original ambitious objective—to design an *automatic* (*algorithmic*) solution—cannot be achieved.

This conclusion is very intuitively understandable: It is not for nothing that a person who follows the same ("algorithmic") pattern of behavior is usually called primitive and dumb. Medieval legends about Golem and other folk "robots" and modern science fiction robot books are full of stories in which the algorithmic dumbness of an otherwise super-smart robot is easily overcome by the creativity ("non-algorithmic" behavior) of a human being.

Since our original problem cannot be solved, we can try to solve a slightly less ambitious problem: to design a *methodology*, a *strategy*, a kind of an "algorithm" in which in some states the next step is *not* uniquely predetermined, but must be chosen from a (predetermined) list of possible steps. In this case, with the (creative) experts making the proper choices, we may hope to solve the original problem.

In short, for intractable problems, we must design a "modified algorithm" that would allow us some *freedom of choice*. Maslov called such "modified algorithms" *calculi* ([**49**], [**52**]).

In these terms, the question becomes: *given a problem, to find a methodology (strategy) for solving this problem*. The question is: how? Maslov has shown that the same idea of freedom of choice that underlies the notion of a "methodology" can actually lead to reasonable methodologies for solving important intractable problems.

1.2. Freedom of choice leads to a reasonable methodology of problem solving. The idea of using freedom of choice to solve problems was developed by Maslov in [**49**] and [**53**] (see also his monograph [**52**]). This idea is easy to explain: Initially, we have a large search space, whose size grows exponentially with the length of the input. For example, for propositional satisfiability with n Boolean variables $x_1, \ldots, x_n$, this search space includes 2^n possible combinations of "true" and "false" values. Because of the huge size of this space, we cannot test all its elements. Instead, we must test only a few "most probable" candidates for a solution. For example, for propositional formulas, we can cut the size of the search space in half if we fix a value of one of the Boolean variables x_i to a certain value ε_i ("true" or "false").

Since we are not testing all the elements of the search space, we may miss a solution. So, we must select a subclass with the smallest "probability" of losing a solution. In particular, for propositional satisfiability, we must select a variable x_i and a value ε_i for which the probability of losing the solution is the smallest possible. After each choice (x_i, ε_i), there may be several solutions.

If we knew exactly the number of solutions $N(x_i, \varepsilon_i)$ left after each choice, then we could simply take a solution for which $N(x_i, \varepsilon_i) > 0$. In reality, however, we do not know these values $N(x_i, \varepsilon_i)$. At best, we know *estimates* $\tilde{N}(x_i, \varepsilon_i)$ for these numbers.

Usually, we have no information about the errors $\tilde{N}(x_i, \varepsilon_i) - N(x_i, \varepsilon_i)$ of these estimates. Therefore, it is natural to assume that larger values of error are less probable than smaller ones. Hence, the larger the estimate $\tilde{N}(x_i, \varepsilon_i)$, the smaller the probability that for this choice (x_i, ε_i) the actual number of solutions will be positive, and therefore, that we will not miss a solution.

As a result, a reasonable method is to look for a choice (x_i, ε_i) after which the estimated number of solutions $\tilde{N}(x_i, \varepsilon_i)$ is the largest possible. In other words, we must make a choice after which *the remaining freedom of choice is the largest possible*. Maslov called this idea "the strategy of increasing the freedom of choice" ([**52**], [**53**]).

As a particular case of his methodology, in 1981, Maslov developed an iterative method for solving propositional satisfiability [**50**] (see also [**51**] and [**52**]).

1.3. Maslov's iterative method: a neural-motivated example of the "freedom of choice" methodology.

1.3.1. *Maslov's choice of a test problem: propositional satisfiability.* By definition, a problem is intractable if every method of solving this problem (in reasonable time) leads to a method of solving all other problems from the class NP (also in reasonable time), So, whichever NP-hard problem we choose, any reasonable heuristic that helps us solve important particular cases of this problem will thus solve particular cases of other intractable problems. Thus, in the long run, it does not make much difference which intractable problem we choose.

In view of this indifference, as a first testbed for his methodology, Maslov simply chose the problem first proven to be intractable: propositional satisfiability.

1.3.2. *Neurons: a source of Maslov's heuristic idea.* In order to estimate the number of solutions, we can try to simulate the way we humans solve complicated problems. Since inside the brain, the processing is done by neurons, it is natural to simulate neurons.

A propositional formula of the type $a \vee b \vee c$ can be reformulated in the form $\neg a \& \neg b \to c$.

- From the viewpoint of the *freedom of choice* strategy, this means that if, according to our estimate, there are many solutions for which $\neg a$ and $\neg b$ are true, then the estimate for the number of solutions for which c is true must also increase.

- In *neural* terms, if we assign a neuron to each literal (i.e., to each variable and to each negation $\neg x_i$), this means that activation of $\neg a$ and $\neg b$ leads to an activation of c and, correspondingly, to a de-activation of $\neg c$.

A natural formalization of this idea leads exactly to the iterative method described in this book.

Historical comment. S. Maslov presented this heuristical neural derivation of his iterative method in numerous talks, but he never published it. The details of Maslov's derivation were published in [**63**].

1.3.3. *Experimental testing of Maslov's iterative method.* In [**52**], [**53**], Maslov tested his methods both on known classes of propositional formulas and on random formulas. The results turned out to be very promising and successful.

The interest caused by this success led to the research summarized in this book.

2. 1981–87: Justifications and modifications of Maslov's iterative method

This book was originally published (in Russian) in 1987. In this section, we will briefly recall the results of this book that later on led to the further research related to Maslov's method.

2.1. Justifications: numerical optimization and uncertainty reasoning.

2.1.1. *Why justification.* In Maslov's original papers, only a *heuristic* justification of Maslov's iterative method was given. The empirical success of this method has shown that this heuristic choice is right (or at least almost right), and that, therefore, a mathematical justification of this method is *possible*.

Do we *need* such a justification? Yes:
- First, a mathematical justification often enables us to *prove* at least some results about the method, and not just rely on its empirical success.
- Second, a mathematical justification often reveals that the method that we are trying to justify is not exactly "optimal" (in some reasonable sense) but only *close* to the optimal method, and thus, enables us to *improve* the original method by finding the truly optimal one.

2.1.2. *Mathematical justifications presented in this book.* In the present book, two main sets of justifications have been presented:
- Justifications based on *numerical optimization* techniques: In [66], M. Zakharevich has shown (see also [29]) that Maslov's iterative method can be obtained as a natural discrete analogue of the simplest and the most well known numerical optimization technique: namely, of the *gradient* descent *method*.
- Justifications based on *uncertainty reasoning*: Numerical values used in Maslov's iterative method can be viewed as describing our current *degrees of belief* in different literals, and the iterative method itself can be viewed as a method of *updating* these degrees of belief. In these terms, Maslov's interpretation of the rule $\neg a \& \neg b \to c$ corresponds to a certain interpretation of an "and"-operation.

  In expert systems, different "and" operations with degrees of belief have been proposed. In principle, it is possible to use these operations (instead of Maslov's original ones) for update; however, empirically, operations that correspond to Maslov's method turn out to be the best. In [28] and [29], this "optimality" is explained by the fact that operations corresponding to Maslov's method are the only ones that have several reasonable symmetries.

The fact that two different mathematical schemes lead to the same iterative method is, in itself, an additional justification that this method is reasonable.

2.2. Modifications: chemical computations and discrete optimization.

2.2.1. *Why modification.* Since Maslov's iterative method is a particular case of the freedom of choice strategy, its empirical success is at the same time the success of the freedom of choice strategy.

It is, therefore, natural to try to expand this success in two directions:
- First, Maslov's method is the result of one heuristic. Maybe, *other heuristics* will be as helpful.
- Second, Maslov's method has been designed for propositional satisfiability. Of course, as we have already mentioned, this means that we can (indirectly) apply this method to an arbitrary intractable problem, because we can always:
  - translate this problem into an equivalent satisfiability problem, and then

- apply Maslov's method to the result of this translation.

But it is definitely desirable to have more direct methods *for other intractable problems.*

Let us briefly describe some of the results from this book.

2.2.2. *Other heuristics: chemical computing.* Several other heuristics are presented in the present book, the most important being the heuristics of *chemical computing* proposed by Matiyasevich in [**54**]. The main idea of this heuristic is similar to Maslov's neural heuristic:
- On the *cell* level, all the data processing in the brain is done by *neurons*. This idea leads to Maslov's heuristics.
- However, we can consider the same data processing on a lower level: From the *biochemical* viewpoint, all the data processing in the brain is performed by *chemical reactions*, so we can simulate chemical reactions and thus get another heuristic. This heuristic is described in [**54**].

The equations presented in [**54**] use chemical kinetics of low concentrations and slow reactions. In [**29**], it is shown that chemical kinetics of fast reactions leads ... exactly to Maslov's iterative method. This additional justification of Maslov's method can be viewed as an additional argument that this method is reasonable.

Matiyasevich has also noticed that in principle, there is no need to *simulate* the chemical reactions on the computer. We can actually *use the actual chemical reactions* to produce *chemical computations* (or *computations in vitro*).

This conclusion is extremely important for computing: Indeed, the faster the computer, the larger problems we can solve in the same amount of time. The speed of computers is currently mainly limited by the speed of light: the smaller the computer's components, the faster the signal passes through them, and the faster the computations are performed. Designers try to make computer processing units that consist of the smallest possible number of molecules. Ideally, we should reduce the size of the unit to a single molecule. In this case, the interaction between these units is what chemists call a chemical reaction. So, *chemical computing is a natural next step in computer design.*

2.2.3. *Heuristics for other intractable problems: discrete optimization.* Maslov's original method was proposed for propositional satisfiability problem. This problem can be reformulated in the following terms:
- we have n sets of alternatives $A_i = \{x_i, \neg x_i\}$;
- from each set, we must choose one of the alternatives $a_i \in A_i$ so that the resulting set of chosen alternatives $(a_1, \ldots, a_n)$ satisfies some given condition.

This reformulation can be naturally extended to the case when we may have more than two alternatives in each class:
- In *graph coloring*, for each of n edges, we must pick a color so that no two neighbors are of the same color. For this problem, A_i is the set of all colors.
- In *job assignment* problems, each of n job applicants must be assigned a job in such a way that each position is taken by only one person. In this case, A_i is the set of all positions for which the ith applicant is qualified.

There are many other natural problems of this type. In [**5**], G. Davydov and I. Davydova have shown that Maslov's method can be naturally generalized to such problems (see also [**29**]).

This generalization led to a reasonably successful method of solving such problems. However, it turned out that the most successful consequence of this generalization is not so much the new iterative method, but rather the new notion of *duality* that became possible only with this generalization. Namely, let us assume that a propositional formula F in conjunctive normal form is not satisfiable. Conjunctive normal form means that this formula is of the type $D_1 \& \ldots \& D_k$, where each of the subformulas D_j is a disjunction of the type $a \vee \ldots \vee c$, and $a, \ldots, c$ are literals (i.e., variables or their negations). The fact that F is not satisfiable means that however we choose a Boolean vector (i.e., however, for every i, we choose an element (x_i or $\neg x_i$) from each of the sets $A_i = \{x_i, \neg x_i\}$), one of the disjunctions D_j will be false, i.e., all its literals $a, \ldots, c$ will be false. We can express this fact in a combinatorial form, if we use the following notation:
- by A we denote the set of all literals;
- by B we denote the collection of sets $A_1, \ldots, A_n$, and
- by D we denote the collection of sets $\{\neg a, \ldots, \neg c\}$ that correspond to different disjunctions D_j.

In terms of this notation, the fact that F is not satisfiable means the following:

For every set $\alpha \subseteq A$, if α has a nonzero intersection with all sets from B, then α must contain a set from D.

Davydovs call a pair (B, D) that satisfies this property *dual*. The interesting property of this notion, a property that could not be noticed before Davydov's generalization of satisfiability, is that duality thus defined is *symmetric*: the pair (B, D) is dual iff the pair (D, B) is dual.

This notion of duality helps us to solve many discrete problems, to cut the exhaustive search, and to develop new non-tree-like search algorithms.

3. 1987–96: Further research

3.1. Applications to real-life problems. The main objective of Maslov's method was to solve *real-life* problems. However, before 1987, this method was only applied to *toy* problems and to *random* propositional formulas.

The well-known 1987 result of P. W. Purdom and C. A. Brown [58] has shown that for several reasonable probabilistic distributions on the set of all formulas, a simple backtracking algorithm requires, on average, polynomial time. This result means that on *random* formulas, all methods work more or less OK. So, to check the actual quality of a method, we must apply it to real-life problems.

In 1987, S. Kamat [25] applied Maslov's iterative method to the following testing problem from computer engineering (to be more precise, he applied Maslov's method to a propositional reformulation of this testing problem):
- We know the results of testing a memory chip. These results do not describe where individual faults are; they describe which functions (involving several memory elements) lead to wrong results.
- Based on these testing results, we must locate the faults.

For this problem, Maslov's methods worked better than the other methods that Kamat tried.

A word of warning. Propositional formulas generated by these testing problems are of very special type. Therefore, the success of Maslov's original method does not necessarily mean that this method will work as well on other classes of formulas. For other classes, other methods (e.g., modifications of Maslov's method) may turn out to be better.

3.2. Applications of the justifications of Maslov's method presented in the book (numerical optimization and uncertainty reasoning).

3.2.1. *Applications of the numerical optimization justification.* In [66], M. Zakharevich has shown that Maslov's method is similar to the simplest numerical optimization technique: the gradient descent method.

Gradient descent is one of the simplest optimization techniques; so if it converges in a few iterations, its running time is small. However, due to its simplicity, gradient descent it not a perfect method: sometimes it converges too slowly; sometimes, it does not converge at all. In numerical optimization, several other methods have been developed. For such methods, a single iteration is usually more complicated, but the total number of iterations is smaller. So, for complicated optimization problems (for which the gradient method fails), these methods are better.

Maslov's original method also often converges too slowly, or not at all. In view of this analogy, for such cases we can try to use analogues of more complicated numerical optimization techniques.

Such analogues were developed and successfully tested by Zakharevich [15], [67], [68]. In particular, he has described a successful discrete analogue of Karmarkar's ellipsoid method (the famous polynomial-time algorithm for solving linear programming problems).

3.2.2. *Applications of the justification based on the symmetry approach to uncertainty.* In [29], a symmetry-based axiomatization was developed for operations on degrees of belief, and operations were chosen on the basis of this axiomatization. This symmetry-based approach to describing and processing uncertainty has since been applied

- to *expert systems* [27], [30], [31], [26], [32], [38], [56], [57]; in particular, this approach was applied: to the design of the *expert system interface* [39];
- to *intelligent measuring instruments* [45], [33], [46], [44];
- to *intelligent signal processing* [34];
- to *intelligent control* [42], [40], [43], [55], [59], [48], [60], [61], [3]; this general "optimization" approach turned out to be consistent with more specific criteria of choosing an uncertainty representation, such as:
  - *stability* and *smoothness* of the resulting control [40], [43], [64];
  - computational *simplicity* [48];
  - *robustness* (i.e., the smallest sensitivity to errors in inputs) [55];
  - maximum *entropy* of the resulting uncertainty [59], [60], [61];
  - and other reasonable optimality criteria [3].

3.3. Further research related to the modifications of Maslov's method presented in the book (chemical computing and discrete optimization).

3.3.1. *Chemical computing after 1987.* The original method of Matiyasevich